MUSCLE CAR COLOR HISTORY

FIREBIRD & TRANS AM
1967-1994

Bill Holder and Phillip Kunz

First published in 1993 by Motorbooks International Publishers & Wholesalers, PO Box 2, 729 Prospect Avenue, Osceola, WI 54020 USA

Motorbooks International books are also available at discounts in bulk quantity for industrial or sales-promotional use. For details write to Special Sales Manager at the Publisher's address

Library of Congress Cataloging-in-Publication Data
Holder, William G.
  Firebird 1967-1994 / Bill Holder, Phillip Kunz.
    p.   cm. — (Motorbooks International muscle car color history)
  Includes index.
  ISBN 0-87938-799-8
  1. Firebird automobile—History.
  I. Kunz, Phillip. II. Title. III. Series.
TL215.F57H65   1993
629.222'2—dc20   93-24761

Printed and bound in Singapore by PH Productions

On the cover: Second and third generation Trans Ams. The red 1986 in the foreground belongs to Tim Willcoxen. Tammy and Bruce Augspurger's 1979 Tenth Anniversary model is in the background.

On the frontispiece: Two generations of the Firebird's top models. The 1971 Trans Am is owned by Dwight Stump of Ohio, and the 1988 Firebird Formula is owned by Gina Clingman.

On the title page: Steve Hamilton's Indianapolis 500 pace cars from 1980 and 1989.

On the back cover: Pontiac's top performance car, the Trans Am, was introduced in 1969. The pristine example shown here is owned by Steve Hamilton.

Acknowledgments

Our thanks to The National Trans Am Club and President Bill Hale; General Motors Corporation, Pontiac Motor Division, Public Relations Department; Deke Houlgate, Pennzoil Corporation, Public Relations; Dick Krieger, consultation; Camaro/Firebird Club; John McKarns, ARTGO Racing; Becky Bayes, National Trans Am Club; Rick Asher, General Motors Corporation, Pontiac Motor Division, Media Relations Department, and Steve Hamilton.

Contents

Introduction

The motivations for the Firebird were countless. Of course, the overpowering reason was the birth of the unbelievably successful Ford Mustang, which proved beyond the shadow of a doubt that the concept of a sports car with universal appeal was a viable one. The opposition had gotten a huge head start, and it was a time for action at Pontiac.

The top management at Pontiac certainly didn't like their faces being rubbed in the dirt by Ford Motor Company's pony. After all, it was GM's Pontiac division that had introduced the first true muscle car, the GTO, in 1964. John DeLorean and Semon E. "Bunkie" Knudsen, the Pontiac big guns at the time, turned their attention to building a Mustang competitor.

General Motors' initial prototype in the "Mustang fighter" category was Chevrolet's "XP-836 Project." It was an effort that began only four months after the Mustang's 1964 introduction. The initial design was a four-seater, a concept that Pontiac had very little interest in adopting.

DeLorean made it very clear that he wanted a two-seater, possibly to serve as a more economic competitor to the Corvette. Thus was born the XP-833 plastic body design. This was a sleek beauty, far ahead of its time design-wise, and in retrospect, maybe just a little too far. The top brass just weren't buying the futuristic concept. Possibly the biggest

reason for the failure of the 833 was the fact that General Motors simply didn't want any competition for the Corvette. End of project.

That's when the Chevy design came back into play as a joint development for both Chevy and Pontiac. Thus it was inevitable that as the two designs evolved there would be great similarity between them—which was certainly the case for many years to come. Ultimately, it just made good economic sense to handle the project this way. Pontiac didn't like the decision, but that was the way the game would be played.

Pontiac's pride was further wounded because much of the design work had already been done by Chevy, and the car had already been given its Camaro moniker. Initially, the name had been Panther, but Chevy decided that it would continue its established pattern of using the "c" names for its cars.

The Firebird name wasn't a new one to the Pontiac division. In 1954, Pontiac built a one-of-a-kind turbine concept car sporting a high rear stabilization bar emblazoned with an early version of the Firebird logo. It looked more like a jet fighter than a ground vehicle. This early design effort was the origin of the Firebird name.

When Pontiac joined the effort, the development had already progressed to a point that the sheet metal contouring was already cast in stone. No doubt, the new

Pontiac machine would have to have its own style and character to set it apart from the Camaro. But with the design work so far advanced, Pontiac had to define its own look using only nose and tail modifications. This was not an easy task.

The same fenders were retained for both models, but the distinctive split-grille design, made famous by the supersuccessful GTO, would definitely set this apart as a PMD (Pontiac Motor Division) design. The Pontiac designers then moved to the car's rear and made their distinct imprint there, too. Even though there were great similarities between the two models when introduced in 1967, each managed to have its own distinctive style—and its own group of devotees.

The Firebird faced a distinct disadvantage against its Camaro brother in that the Camaro was released much earlier than the Firebird. Camaros were released to the public in September 1966 and were carried as 1967 models. Firebirds didn't make it into showrooms until February 1967. Consequently, it was labeled a "1967-1/2" model.

Quite frankly, force-fitting the Firebird within already-existing design constraints had left it way behind the Camaro. Still, the Firebird line has survived through the years and has continued to forge its own personality. No longer a Camaro hanger-on, the Firebird (and its flashy Trans Am offshoot) has moved in a magnificent direction.

This was the prototype for the first Firebird, denoted as the "GM-X" by the emblem fitted to the front quarter. Much of the styling of *this prototype was retained in the initial, first-generation Firebird design.* Pontiac Motor Division

The GM-X prototype design introduced the distinctive rear-end look for the first Firebird. Even though the design was con- *strained by having to use existing Chevrolet Camaro sheet metal, the Firebird still had its own style.* Pontiac Motor Division

The purpose of this book is to trace the changes and modifications and chart the direction the Firebird breed has followed through the years.

Even though the Trans Am is the top model of the Firebird line, many look upon it as a completely separate model. For that reason, the Trans Am will be addressed separately in this book. Most owners of this classic line will tell you that they have either a Firebird or a Trans Am. Enough said! All the great high-performance engines that have powered these cars through the years will also be reviewed. Finally, we'll take a brief look at the great racing history of both models.

During the current vintage muscle car rage, Firebirds seem to be the forgotten muscle cars. But make no mistake, many versions of both the Firebird and Trans Am more than qualify as muscle machines. In fact, from 1973 to 1974, Firebirds and Trans Ams were among the few muscle cars still on the market.

Although the Super Duty 455 powerplant of those dark EPA years was the final gasp of the glorious muscle era, Firebirds have survived into the 1990s and still offer some of the most exciting performance driving available.

7

Underhood Performance

Through the 1960s and early 1970s, the Firebird and Trans Am lines were endowed with some of the most powerful engines of the muscle car era. Modern Firebirds are no slouches either; and thanks to current technology, they possess performance rivaling the early Firebirds. But it's the muscle era engines that solidly classify early Firebirds and Trans Ams as desirable muscle cars.

Performance was evident even in the first base-model Firebird. The Pontiac-designed inline six-cylinder displaced 230ci and boasted an overhead cam. Outfitted with the optional Sprint package of hotter cam and four barrel carb, the six made an impressive 215hp.

The initial V-8 powerplant was a 326ci mill (in the sporty HO version) which featured a four-barrel carb, dual exhausts, and a 10.5:1 compression ratio. Its 285hp output was a portent of things to come.

The Super Duty SD-455 engine was a powerhouse in a period of bland, no-punch engines. It was introduced in 1973 but stayed in the lineup through just 1974. Its advertised power of 290bhp was really a joke, as proved by its 13sec performance in the quarter-mile. The engine was packed with plenty of high-performance innards, and there's really been nothing like it since.

For the 1970 model year, the top powerplants offered for the Firebird were the 400ci Ram Air III and IV. The Ram Air III (shown here) produced a ground-pounding 330bhp (some factory literature claimed 335bhp). The Ram Air IV produced 345bhp and sported a 10.75:1 compression ratio. Both carried the distinctive Shaker Hood and a Rochester four-barrel carburetor.

Bigger cubic inches and higher horsepower were also available to those early Firebirds, including two versions of the vaunted 325hp 400ci engine (the Ram Air version achieved its peak power at a higher rpm). A 3000lb car with that kind of power equalled serious performance!

A year later, in 1968, a 350HO (high output) engine producing an impressive 320hp was made available. If this wasn't sufficient, one could order a ground-pounding 330hp 400ci powerplant.

In 1969, the horsepower numbers rose to 345 with the 400ci Ram Air IV powerplant. This was also the first year for the fashionable Trans Am model, which would share the potent powerplant. A Ram Air III (400ci HO version) was available with ten fewer ponies.

Power remained unchanged for 1970, but in 1971 the cubic inch options expanded to include a new 455ci big block. Interestingly, even with the additional cubic inches, the horsepower numbers remained the same as the previous standard 400ci and Ram Air III ratings of 325 and 335 horsepower. However, the 455's torque was greater.

Horsepower and compression ratios then started their decline, with one major exception occurring in 1973 and 1974. It was called the SD-455 Super Duty powerplant, and it was a killer! Pontiac listed the horsepower at a conservative 290, but everybody knew that was an understatement. The NHRA factored the engine at 375hp, which was probably closer to the actual value. Collectors consider this powerplant as possibly the most desirable of the vintage high-performance Pontiac engines.

For a number of years during the 1970s and 1980s, the term "high-performance" was not in Detroit engineers' dictionaries. But in the late 1980s, the words reappeared in the Firebird showroom sales brochures.

In 1987, for example, it was possible to get 210 net horsepower (horsepower measured with all the accessories attached and operating) from a 350ci powerplant. In 1988, the numbers went to 235 net horses from a 5.7 liter Tuned Port Injection (TPI) powerplant. A Special Edition Pace Car version in 1989 carried a 3.8 liter turbocharged version which was worth 245 horses. A TPI 5.0 liter engine was the top gun for 1991 with 230 horsepower.

From all current trends, it appears that performance will continue to play a leading role during the 1990s. It also appears that the Firebird and Trans Am will continue to be the recipients of some of that capability. That's the way it's always been, even though these Pontiac pony cars have never gotten the "muscle" recognition they justly deserve.

The 455 was new for 1971 with the 455HO powerplant being the top performer. The engine was identified by a small "455HO" decal on the Shaker Hood. Power was advertised as 335bhp—pretty impressive considering that the compression ratio had been dropped to 8.4:1.

Pontiac Motor Division

After this, you'll never go back to driving whatever you're driving.

If you can stop drooling for a moment, we'd like to tell you what's propelling that Firebird 400 in the picture. What it is, is 400 cubes of chromed V-8. And what it puts out is 325 hp. (Even without our extra-cost Ram Air package, that makes those dual scoops functional.)

The point being, that Pontiac Firebird 400 was designed for heroic driving.

To assist you in this noble venture, the 400 comes with a heavy-duty 3-speed floor shift, extra sticky sus-pension and a set of duals that announce your coming like the brass section of the New York Philharmonic.

Taken as she comes, Firebird 400 is a lot of machine, but you can order things like a 4-speed (or our stupen-dous 1-2-3 Turbo Hydra-Matic), mag-type steel wheels, special Koni adjustable shocks and a hood-mounted tach. Naturally, the GM safety package is standard.

Of course, if the 400 is too much car for you, there are four other Firebirds to choose from. Lucky you.

Firebird 400. One of Pontiac's Magnificent Five.

Picture this. We'll send you six 24" x 13¾" full-color pictures of Firebird 400, Pontiac 2 + 2, GTO and OHC Sprint, plus complete specs and decals. Send 25¢ (35¢ outside USA) to '67 Wide-Tracks, P.O. Box 880E, 196 Wide-Track Blvd., Pontiac, Mich. 48056. Include your ZIP code.

Firebird 1967–1976

1967-1/2 Firebird

As the first year model, the 1967-1/2 Firebird is considered to be one of the most valuable of the "Bird" breed. For a first-year car, there was an amazing selection of models and motors. The public obviously liked what it saw, buying some 82,560, of which 15,528 were convertibles. It was a significant sales accomplishment, considering that the model had a monstrously late February 1967 introduction.

Pontiac used its generous option list in an advertising blitz, asking potential buyers: "Which Firebird is for You?" Pontiac tried its best to divorce the model from its obvious Camaro heritage, and marketed the Firebird as a distinctive, unique sports car.

When viewing the model from the side, its Camaro resemblance was obvious due in no small part to the fact that the Camaro's front fenders were retained (a GM requirement). But PMD engineers had established an elegant appearance in the front and rear treatment. There was a Pontiac GTO-style hood, with horizon-

tally mounted twin-headlights recessed within the distinctive blacked-out hood. At the car's rear were four horizontal taillights (two at each corner). The Pontiac identification was carried in the lower center of the rear deck.

Five models were available the first year: the Firebird, the Firebird Sprint, the

The first ads for the new Firebird accentuated its dazzling looks and 325hp performance. Although the Firebird was a late-year production, it had a popular inauguration.

The "HO" letters embedded in the side-trim stripes identified the punchy 326HO model. This model carried the High Output, 285hp version of the 326ci powerplant. There were

also heavy-duty suspension and a column-mounted shifter included in the $180.58 package.

The baseline powerplant for the first Fire-bird was a six-cylinder 3.8 liter mill with both a single- and four-barrel carburetor available. Horsepower was expectantly quite different, with 115hp and 215hp levels available. But if you wanted performance this in-augural year, the 326ci and 400ci power-plants were the way to go.

Previous page
Although the design of the initial Firebird was forced to conform to the sheet-metal shaping of the Camaro, the Firebird still had a charisma of its own. Its distinctive split-grille design would remain for many years and become its trademark. The design still retains a classic look in the 1990s.

Firebird 326, the Firebird 326HO, and the Firebird 400.

The $2,600 base Firebird coupe, was a bare bones model, but it could be upgraded to a ragtop version for an additional $237.

The base Firebird certainly couldn't be confused with the muscle cars of the era. The model carried a 230ci six-cylinder topped with only a single-barrel carburetor. The 165hp rating was minimal compared to other optional engines in the family and just over half that of the top-of-the-line Firebird 400 model. The compression ratio was a low-for-the-period 9:1.

The emblem to go with the new Firebird name had to be special, and what PMD came up with was exactly that! The classy-looking bird has lasted through the years while numerous other features on the cars were changing with the times. The bird truly sets the car apart from its Camaro running mate.

The Firebird Sprint was the next step up the line, hitting the wallet for an additional $116. It was as though the base powerplant had been dropped off at the neighborhood speed shop for the Sprint application. A hotter cam, split exhaust manifold, low restriction air cleaner, and a Rochester four-barrel carb added an additional 50hp to the Sprint. The high-revving little powerplant also perked at an impressive 10.5:1 compression ratio.

There were several appearance and performance modifications made to set the Sprint apart from the base model. A floor-mounted three-speed shifter and firmer suspension made the driver feel that he had a formidable performer on his hands. The Sprint, which weighed 55lb more than the base model, also carried "OHC-6" emblems on the lower body

rocker panels. It wasn't muscle car punch by any means, but its performance was impressive, especially from a little in-line six.

Firebird 326ci models moved the engine displacement up 94ci, but increased horsepower by only thirty-five over the Sprint six. Despite being a somewhat laid-back V-8 powerplant—a 9.2:1 compression ratio and a one-barrel carb—it made an impressive 333lb-ft of torque at only 2800rpm. Even though the 326 possessed more horses than the Sprint, the model lacked the Sprint's pizzazz.

The Firebird 326's shifter was column-mounted, and the engine was actually twenty-one dollars cheaper than the Sprint's. It came down to a different design philosophy for each model, as well as a different driving style. Identification

of this model included "326" numerals on the hood.

The Firebird 326HO got the buyer back to performance in a big way! There was a 326ci high-performance engine under the hood that made this model a real rocket and one of the most desirable of the early Firebirds.

The $280 HO Option featured an impressive 326ci, 285hp engine that many said was totally underrated. Several Firebird experts have indicated that they felt the figure was actually more in the 300hp, or greater, range.

The HO produced 359lb-ft of torque and included such performance goodies as a Carter four-barrel carb, dual exhausts, and a 10.5:1 compression ratio.

The HO body was distinctively signified with a body-length horizontal stripe

The 326 High Output powerplant was one tough little mill, punching out 285bhp at 5000rpm. The engine carried a four-barrel carb, dual exhausts, heavy-duty battery, and sported a 10.5:1 compression ratio. A first-year Firebird carrying this engine is a desirable collectible.

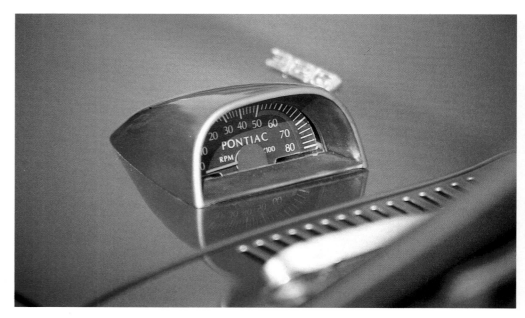

The optional, hood-mounted tachometer was far ahead of the competition at the time of its introduction on the first Firebirds. Mounted on the driver's side of the hood, it was easily visible from the pilot seat. With an 8000rpm dial, the tach was a $63.19 option.

The hot performance number for the 1967 Firebird line was the Firebird 400, which carried an awesome 325hp 400ci power-plant under the hood. The model featured the flashy twin hood scoops, dual exhausts, heavy-duty three-speed transmission, and sport suspension. The package cost the buyer an extra $274.

with an embedded "HO" identification. Production of 326HOs was very low, but exact numbers are uncertain. Many lived a brutal, and possibly short, life on the nation's drag strips and may not have survived to the present.

But if you wanted the top dog performance-wise of the first year Firebirds, the Firebird 400 was the ticket. There were actually two versions of this model, even though both were rated at a punchy 325 horses. The Ram Air version, which added an additional $263 to the bottom line, featured functional hood scoops (nonfunctional on the standard 400 engine) which enabled the advertised horsepower to be accomplished at lower revs than with the base 400 version.

The top 400 powerplant was very similar to the vaunted GTO powerplant, but it carried a lower lift and shorter duration cam. The torque of both Firebird 400 engines was a stump-pulling 410lb-ft.

Other options on the Firebird 400 included a floor-mounted three-speed transmission, dual exhausts, and a heavy-duty sport suspension.

As with any new model, there were certain teething problems with the first Firebirds, such as a tendency to oversteer under heavy load conditions. But overall, the response was enthusiastic to the "Camaro's cousin."

Despite the disadvantages of a late-year introduction and force-fit engineer-

The interior of the first Firebird certainly lived up to the car's exterior looks. Definitely classified as a sports car, the 1967 Fire-bird could be decked out with bucket seats, console and floor shifter (a $47.39 option), deluxe steering wheel, and many other items.

The convertible model shown here is one of the most desirable collectibles of the initial Firebird run.

National exposure was important to the introduction of the new Firebird, and as has been the case many times hence, racing provided a needed venue. This autographed 1967 Firebird was the first Firebird to be a pace car, the duty being performed at the Daytona International Speedway.

ing within an existing design, Pontiac still managed to produce a significant and highly flexible machine. It was a

great start for the Firebird, but the best was yet to come in both looks and performance.

1968 Firebird

To borrow a cliche, if it ain't broke don't fix it! That proved to be Pontiac's thinking for the 1968 Firebird. But yearly model changes were the status quo in the 1960s, so there were some "minor refinements" made. The philosophy seemed to be right on the money because 107,112

units were sold, a number that wouldn't be exceeded until 1976.

Nineteen sixty-eight was a year of refinement, and if you didn't look carefully, it was difficult to detect the external differences between the two model years. Let's consider the sheet metal changes first.

The Firebird seemed to be moving into the modern era with the elimination of the old-style vent windows. This change gave both a sleeker overall look

21

Front-end changes to the second Firebird were minimal—why mess with success? The parking lights were removed from the lower grille slots and wrapped around the front body corners. The Firebird would also show side-marker lights on the rear flanks for the first time. Just a few little refinements to make the great even greater.

to the body and provided improved ventilation.

Bodyside marker lights (in the shape of the PMD medallion) appeared on the rear quarters. The "Firebird" lettering with the Bird symbol spread its wings

just forward of the name tag, identical to the 1967 treatment.

Open the doors, and bigger changes were in evidence. Liberal use of attractive, burled woodgrain detailing graced the dash and console. Vinyl inserts were added to the seats to change their styling. A nice performance option was a hood-mounted tach.

A wheel hop problem, which had received some bad press on the initial Firebird, got considerable attention on Firebird number two. To address that problem, multi-leaf springs and staggered shocks were installed on all but the base model.

The base powerplant was again a single-barrel-carbureted six-cylinder. Displacement grew to 250ci, up twenty from the previous year, while horsepower showed an increase of ten, to 175. Torque was measured at 240lb-ft, while the compression ratio remained at 9:1.

Only minor performance refinements were made to the in-line six Sprint engine. Interestingly, the horsepower would remain the same, at 215, but the torque was up slightly to 255lb-ft.

As in the previous year, the $116 Sprint package included a number of trim items and a floor shifter in addition to the higher performance powerplant.

The rear of the 1968 Firebird again carried the stylish bird emblazoned on the gas cap cover. Beneath the cover, the second Firebird sported a stronger rear bumper, but the overall design was almost identical to the previous model's.

The new Firebird 350 package came next, and carried a very unromantic (and unsports-car-like) column-mounted three-speed shifter. Bucking the performance trend at the time, this model's engine was fitted with only a two-barrel carb. Still, there was an impressive 265hp at 4600rpm on tap. Despite its larger engine, the package was still ten dollars cheaper than the Sprint package.

Like 1967's 326, the 350 was available in an "HO" version and carried the same distinctive stripe, only with a "350" embedded. The shifter was still on the column, and performance goodies included dual exhausts, heavy-duty battery, and F70x14in rubber.

The powerplant provided a pleasurable punch with an impressive 320 horses. This was a great number for a small-block, close to the one-horsepower-per-cubic-inch goal of the day. A Rochester four-barrel and 10.5:1 compression ratio were big factors in achieving this engine's rating.

The 350HO has attracted interest in the collection community, but is considered somewhat of a marginal muscle car by some. When new, the machine was a nice performance alternative to the big engine machines for a younger buyer. The 350HO was also very attractively priced at only $181 over the base model.

But any muscle car fanatic of the 1990s will tell you that the 400ci versions (there were three in 1968) were the hot tickets. With the exception of the few 427 COPO Camaros that were produced that model year, Pontiac had a slight step up on the Camaros, as that machine's biggest mill was a 396.

The shifter, moved to the floor, controlled a manual three-speed for the 400ci powered versions. The standard package was the Firebird 400, but to real-

A number of options was available with the stylish interior of the 1968 model. The Custom Trim option featured the wide-vertical-pleat seat shown here. Also available were a deluxe steering wheel, woodgrain console and dash, stereo tape player, rear folding seat, and many other options.

A completely new treatment marked the grille of the 1969 Firebird. The headlights were no longer surrounded by chrome; now, only the twin-opening grille was so encased. Lexan plastic was also used for the first time in this front-end treatment. Whether you liked it or not, there was certainly a different look to the third Firebird.

ly light your fire, there was also a Ram Air version. Again, there was minimal difference power-wise between the pair (only 5hp), with the Ram Air version rated 335hp at 5000rpm.

The Firebird 400 package included flashy chrome detailing on the air cleaner, valve covers, and oil cap; a heavy-duty suspension system; and special model identification emblems.

The standard 400 powerplant pushed 10.75:1 compression and carried a unique power-flex fan and a Rochester four-barrel carb. The Firebird 400 wasn't cheap, though, requiring an extra $435 over the base model.

The top dog in the performance category, was the Ram Air 400 option that carried the desirable twin, functional hood scoops. These were supposedly all-weather scoops that could always remain open, but most owners blocked them off during the winter.

A second version of the Ram Air powerplant, called the Ram Air II, was introduced near the end of the model year. To call this version an upgrade doesn't do it justice, because it was effectively a new powerplant and featured such heavy-duty items as four-bolt main bearings, special manifolds, and forged pistons. Very few of these models were produced. In the years to come, they're sure to bring a high asking price. Pontiac experts suggest that the Ram Air II could be one of the most desirable muscle collectibles of the Firebird breed.

Its second model year, 1968, was a great year for the Firebird. The engines, models, and options seemed to flow without end. Pontiac's sporty offering could answer just about any desire be it from a performance or an appearance perspective.

There were few changes made to the rear end of the 1969 Firebird. The style was characterized by the wide, horizontal twin taillights on each side. Note the dual exhausts on this 400 model. The rear spoiler was a dealer-installed option.

1969 Firebird

A time of transition best describes Firebird's third model year. Name a section of the car, and it was probably changed for this model year.

First of all, there was a complete sheet metal restyling both inside and out. Then the awesome Ram Air III and IV powerplants were introduced. Finally, the formidable Trans Am model debuted. The overall effect was that the Firebird really established itself as a distinct pony car, not just a Camaro copycat. But even with all this positive change, production still experienced a downturn to 87,011, far from 1968's sales bonanza.

As in previous years, Pontiac offered a montage of models and options to perfectly fit even the most discriminating buyer.

As far as the body restyling was concerned, public reaction was either love or hate. The front-end treatment seemed to lose some of its original style. Headlights were now encased in body-colored pods, and the grille looked as if it had been squeezed to the center, as if it were puckered up.

Some of the sheet metal changes followed those made on the Camaro that year, such as a slightly lower profile. The wheel openings were flattened at the top, a noticeable design deviation from the rounder styles of the earlier years. Two different hoods were also available, with an integral scoop version available for the Ram Air powerplants.

New interior designs were another highlight of these last-of-the-decade models. The dash design was all new and made gauge-monitoring easier for the driver. Bucket seats were the vogue and were greatly improved for 1969. A bench seat was a rarity in the Firebird, but who'd want a bench seat with a muscle sports car of this caliber?

Trim options for the interior abounded, there being no such thing as a standard layout. There were also numerous safety features added, including the first appearance on a Firebird of over-shoulder seat belts.

An amazing array of powerplants was offered in 1969, the base engine still being the 250ci inline six (though now making 175hp).

26

The popular Sprint model was rewarded with an impressive 230hp from a powerplant of only 250ci. The formidable little mill featured a Rochester four-barrel carb and a 10.5:1 compression ratio. It cost the purchaser a well-invested $121 over the base model. Sprint identification was carried on the rocker panels and hood.

The Sprint is still a popular number, as demonstrated by a 1991 advertisement in a national car magazine in which a completely restored 1969 model was offered for $10,000!

Add 100ci and two cylinders, but only 35hp, and you have the Firebird 350 model. The performance from this L-30 engine was dictated by the rather low 9.2:1 compression ratio and its two-barrel carb. Surprisingly, it was the most popular sales-wise of all the Firebirds that were built that year, even during this performance era.

Add the "HO" designation to the 350, however, and its true power potential quickly came to light. Known as Option Code 343, the L-76 350ci engine pounded out a big-block-like 325 ponies at 5100rpm. Special internal performance parts included a higher lift cam, larger valves, and new cylinder heads. The HO option was a solid 15sec performer in the quarter and required only an additional $186 over the base model.

But as in years past, if you wanted the ultimate performance, you had to turn to the 400. This year, high performance was selling like hot cakes and there were three versions of that big-block to quench the thirst.

The Firebird 400 was the base model, Option Code 345, and carried the W-66

The 400HO (or Ram Air III) powerplant was the number two powerplant for the Firebird in 1969. The powerplant provided 335bhp and added $77 to the purchase price. The more powerful Ram Air IV with 345bhp was an option with the model.

engine. At 330hp, it provided only five more ponies than the 350HO, but it made considerably more torque at 430 lb-ft, versus 380lb-ft for the 350HO.

Even though the Firebird 400 was the entry-level 400 equipped model, it contained considerable glitz and a number of chromed engine parts. Flashy hood scoops lent the car that high-performance air, even if the scoops were nonfunctional. The shifter was floor-mounted, and dual exhausts protruded from the rear. For many, it was all the performance they needed (or could afford) because it added $358 to the base price. But for those who wanted additional firepower under the hood, there was much more available!

The next step up the performance ladder was the $435 Ram Air 400 package, Option Code 348, which was a refinement to the basic 400 package. The L-74 powerplant used the hood scoops for their intended purposes.

But even with the additional induction the functional hood scoops provided, the horsepower numbers were raised only five beyond the baseline 400's. The torque of the two engines was actually the same, at 340lb-ft.

The ultimate performance option deal for this year was the Ram Air IV. The L-67 powerplant in this package was similar to the W-66 and L-74 mills, but it incorporated valve train changes and a hotter cam. The result was an additional ten horsepower raising the L-67's output to 345 at 5400rpm, and a torque figure (430lb-ft) identical to that of the W-66.

The Ram Air IV, at $832, was monstrously expensive for the time. No wonder few were sold, and even fewer exist today.

Although many Firebird lovers believe the 1969's combination of looks and performance make it the top year for the model, production was off, with just over 87,000 produced. The production figure was actually even lower than it appeared, as the model was sold (be-

There were minor changes to the Firebird design for the 1969 model year. Note the front quarter trim pieces, which were different from the air extractor scoops carried by the Trans Am the same model year.

cause of labor problems) until the end of 1969, or for a total of 15 months. The later models were actually sold with 1970 titles.

Not included in that production figure were 697 of a new Firebird variant called the Trans Am. But that's another story to be covered later in the book. Although the Trans Am started off as a Firebird option, the model would assume its own identity and informally be considered as a separate model in itself. For that reason, the early Trans Ams will be addressed separately.

1970 Firebird

It was a new realm for the Firebird in 1970. Actually, it was a 1970-1/2 situation, because of the late arrival of this completely restyled version. Pontiac called the models "1970+" versions that year.

Just about everything was new: the sheet metal, the engines, and, maybe most significantly, Pontiac's more worldly perception of the model. The new outlook was manifested in the names of the two top Firebirds, the Esprit and the Formula 400. The competition for these machines would be broadened beyond just the Mustang to include the worldwide market.

The PMD designers did their job well, producing a European design incorporating a perfectly integrated Endura (a plastic material) grille and a sleek, fast-back body style. And, for the first time, there would be single headlights.

Change in 1970 went far beyond a new skin, however. There were considerable upgrades to the suspension system, brakes, and steering. There were also beefy stabilizer bars fitted to the Formula 400.

Again, there was sheet metal identical to Camaro's, which also brought the new design into its line. The outer-skin similarity between the two models would continue until 1981.

The 1970 Firebird looked smaller than its earlier incarnation, even though it was actually just over 1in longer and a 1/2in narrower than the 1969 model. The wheelbase was identical, but both the

The Lexan body-colored grille appears to stretch the length of the hood. This particular 400 model carries the Ram Air III pow- *erplant with the displacement denoted twice on the hood sheet metal. The 1969 model would be the last of this particular design,* *one that many Firebird lovers consider the best there ever was.*

The 1970 Firebird was completely restyled and the consensus from the public was a thumbs up. The restyled front end featured single headlights and a new-look Endura body-colored grille. The Formula 400 model shown here featured a fiberglass hood and the stylish scoops. This machine was a looker of the first order.

Pontiac's Esprit model was tagged with a luxury label and it was slotted into the Firebird lineup directly below the top-of-the-line *Trans Am and the Formula. The 255bhp, two-barrel-carb 350ci mill was standard with the Esprit. The Esprit nomenclature* *was mounted on the rear top support. Pontiac Motor Division*

front and rear wheels had been moved forward 3.5in.

Even with the new look, sales for this year's model were horrible—only about half of 1969's total. Had Pontiac made a mistake with the new design, or was it just a reflection of the lagging pony car market? The new Firebird sales effort only attracted 45,543 buyers (not including Trans Ams).

The base Firebird used a 250ci Chevy engine, replacing the previous Pontiac powerplant. Although the displacement

number were identical with the PMD mill, the performance was down considerably, dropping from 175 to 155hp. This decline was due in part to a lower compression ratio (8.5:1).

But there was hope for the base Firebird, because for the first time it was possible to specify a V-8. Even though the bottom 350ci mill had a woefully low 8.8:1 compression ratio and a two-barrel carb, it was still capable of 255hp. That was the same horsepower figure the 1969 L-30 350 had generated, even though

that engine had the advantage of a 9.2:1 compression ratio. This 1970 350 version was denoted by "350" numbers in red on the front fenders.

The Sprint was gone, replaced by the Esprit. Suavity and style, not performance, were this machine's selling points. The Esprit was plush inside, with a new ventilation system, sport steering wheel, and custom knit vinyl seat coverings.

The same 350 that was optional for the base Firebird was the Esprit's stan-

The Ram Air III 400ci powerplant was the less powerful of two Ram Air mills for 1970. This mill sported 335bhp, only 10bhp fewer *than the Ram Air IV version. This engine created 430lb-ft of torque at 3400rpm derived from a four-barrel carb and 10.5:1* *compression ratio. Firebirds equipped with these powerplants are some of the most desirable muscle car collectibles.*

dard engine. A floor shifter directed the 255 horses which had to push 3,435lb, a portly 295lb more than the base model. The Esprit sticker price was $3,241, a $366 increase above the entry level Firebird. Sales totals for the Esprit weren't bad for its first year, coming to 18,961. Those totals, however, were certainly a far cry from the earlier years. Things were changing in the muscle car market.

All three versions of the 400-powered 1969 Firebirds were replaced by the new 1970 Formula 400, with two versions of the 400 mill available. The engines were identical to the base 400 (330hp) and Ram Air III (335hp) of the previous year. A Ram Air IV version with the same performance as 1969's

33

The redone interior for the 1970 Firebird was both functional and beautiful. The racing-style steering wheel was a buyer favorite. The dash was packed with instrumentation surrounded by a simulated panel. The center console blended in with the sports-car look of the exterior.

was also available in 1970, but only as a special order. First year production for the flashy Formula was 7,708, of which 4,931 were automatics.

The Formula had the looks and equipment to go with the trendy name. There were both front and rear stabilizer bars, front disc brakes, and high-rate springs.

Externally, the Formula looked like a sleek road racer, with its sport-type wheels and rakish styling. The long, functional snorkel scoops swept the length of the fiberglass hood and terminated just short of the Endura grille piece. Sport mirrors and concealed wipers helped maintain an aerodynamic appearance. The interior was attractive as well, carrying front and rear bucket seats and carpeting.

The car magazines reported favorably on the new designs, both from appearance and performance points of view. *Hot Rod* magazine tested a 330hp version of the Formula and ran it through the quarter in 15sec flat at 93mph.

1971 Firebird

Nineteen seventy-one was a year of contradiction for the Firebird. Despite an industry-wide decline in the horsepower trend due to new anti-pollution requirements and higher mileage goals, Pontiac chose 1971 as the year to introduce its new big-block 455ci engine. The two just didn't seem to mesh.

The 455 was the largest displacement engine ever installed in a performance model, and the Firebird and Trans Am were both recipients of the giant mill. Interestingly, in a little battle of displacement numbers, the Pontiac powerplant was one up on Chevy's killer 454 engine.

An integral spoiler was a design highlight of the 1970 Firebird. The clean spoiler styling gave it the look of a race car. There were considerable improvements to go with the outside looks, including improved suspension and steering along with front disc brakes now standard.

There were only slight changes made to the 1971 Firebird. The main external appearance change came with the simulated louvers added to the front quarters. There were also upgrades made in the interior. With the base V-8 powerplant, the base price of the model was only $3,168. Pontiac Motor Division

The big news was under the hood for the 1971 Firebird. On the outside were just minor cosmetic deviations from the 1970 models. Still, there were seven different models available, providing the buyer with everything from mild to wild.

The base Firebird was available with either a six-cylinder or a V-8. The six obviously wasn't built for performance, since the 250ci engine was rated only at 145hp (down 10hp from 1970). Such minimal performance attracted only 13 percent of the base Firebird's buyers. It cost an additional $121 to acquire the same 350ci V-8 that powered the T-37 and LeMans models.

The Esprit was again the next step up the model ladder, with two engine options available. Luxury was the keynote, with a custom steering wheel, knit vinyl upholstery, and concealed wipers. Outside, there were body-colored mirrors and wheel trim rings. Considerably beefier than the base model (259lb heavier) at 3,473lb, the Esprit required an additional $369.

The base Esprit carried the 350ci V-8, two-barrel engine which provided 250hp at 4600rpm. The compression ratio was 8.8:1, which was fairly standard for the period. Also becoming standard was the quoting of *net* horsepower (nhp) instead

The Rally II wheel was one of three available wheel options in 1971. The five-spoke design featured the PMD center insert and five chromed nuts. Slight variations of this design would be available for a number of years on the Firebird line.

of the long-standing brake horsepower (bhp).

The net number was considered a more realistic rating of the engine's power while driving all the accessories. Net numbers usually work out to be about 80 percent of the brake horsepower number, thus giving an impression of a greater degradation in horsepower than was actually taking place. In the case of

this particular 350 engine, the 250 brake horsepower related to only 165 net horsepower. During 1971, all horsepower ratings carried both numbers, but in subsequent years only the net numbers would be provided.

A downgraded 400 was also available for the Esprit. With a two-barrel and only 8.2:1 compression ratio, the engine put out 265bhp, or 180nhp.

Finally, there was the Formula, available in three versions. At $3,445, the base-model Formula 350 was priced just $29 more than the Esprit, but it seemed that it was worth a lot more.

There was a wide array of standard equipment, including vinyl bucket seats, woodgrain dash, custom steering wheel, Endura front bumper, handling package,

bright grille mouldings, and simulated hood scoops.

The 400ci powerplant in the Formula 400 was down a bit on power from previous versions, but still potent at 300bhp (250nhp) and 400 lb-ft of torque.

But it was the pair of thundering 455s that really got the public's attention. The less-powerful 455 had a compression ratio of only 8.2:1, but still produced 325bhp (255nhp) and added $100 to the base Formula price. The torque stood at an amazing 455lb-ft, exactly matching the engine's displacement.

Then there was the killer 455HO ($237 extra), which was capable of 335bhp. The powerplant carried a four-barrel, 8.4:1 compression ratio mill controlled by a floor-shifted, three-speed transmission. The 455HO would also be the standard engine for the Trans Am, but made only rare appearances in the Formula. It's a model worth pursuing in the 1990s, owing to its low production numbers and high power. Definitely a sleeper.

In all, there were only 53,124 Firebirds, including Trans Ams, produced in 1971. But it was an interesting year, and deserves closer inspection from a collectible point of view. And it goes without saying that any Firebird carrying that first-year 455 is worth serious consideration.

1972 Firebird

Things did not look good for the Firebird in 1972. The Firebird's production trends told a sad story. From a high of over 107,000 in 1968, Firebird sales showed a straight-line decline to less than half that figure by 1971. Compared to those sterling early sales numbers, it hardly seemed worth firing up the line.

There was even talk, serious talk, of dropping the Firebird line, and the more popular Camaro line as well. High insurance rates along with spiraling gas prices made many turn away from the gas-sucking performance machines.

To control costs after the decision was made to continue production, there were only minimal changes made to the Firebird. The most noticeable difference for all versions was the introduction of a honeycomb mesh grille.

In keeping with the times, the base model (Model 2S) was the best seller at 12,000. Standard equipment for the bottom version were front and rear bucket seats, vinyl trim, loop-pile carpets, a deluxe steering wheel, walnut-detailed dash, and the popular Endura bumper. The base model had a maximum sticker price of $2,956.

The net horsepower was a minimal 110 from the Code D in-line six-cylinder powerplant, the figure listed exclusively in net horsepower as would be the case from that year forward. The engine was basically identical to the 1971 version.

Also available again was the optional Code M two-barrel 350ci V-8 powerplant now down to 160nhp. The compression ratio continued to decline as well, dropping to 8:1. With dual exhausts, the horsepower numbers for the optional Code N 350 picked up fifteen additional ponies.

Add $238 to the base-model's price and the Esprit (Model 2T) came into view. The glitzy model featured script moldings on the roof pillars, as well as a custom steering wheel, wheel trim rings, concealed wipers, wheel opening moldings, and rocker panel accent strips. The sport floor-mounted three-speed was also standard equipment. A total of 11,415 Esprits were sold, with 10,911 equipped with the Hydra-matic automatic transmission. One of the 504 copies built with a stick shift would make a very interesting collectible.

The Esprit 400 carried the L65 400ci powerplant hooked to the M40 Turbo Hydra-matic transmission. With duals, the package was worth an impressive 175nhp. If you find one of these Esprit combinations, it might well be worth considering for purchase.

The Formula was again the top dog (with the exception, of course, of the Trans Am), and it no longer carried the Firebird imprint on its sheet metal.

The dire sales predictions that had preceded the 1972 model year were somewhat manifested, as only 5,250 Formulas were produced. About 20 percent of those (1,082) carried the manual synchromesh transmission.

Depending on the engine choice, the Formulas were identified as the Formula 350, Formula 400, or Formula 455. And even though the performance trend was in a downward direction, the horsepower figures for these models were deceptively low, again because of the "net" ratings. Remember, net figures were actually some 20 percent lower than the equivalent brake horsepower numbers numerically.

Formulas could be ordered with the Code M or N 350 engines (160 and 175nhp, respectively) or with the Code R L-78 400ci, four-barrelled powerplant. The 400 was rated at 250nhp and could certainly stand on its own in the street wars of the time.

The 455 was again the top engine available for the Formula. Known as the LS5, its net horsepower figures were carried at 300, but when converted to the "brake numbers" could well have been in the 330bhp range. There was a solid 14sec quarter-mile performance to back those numbers up.

The 1972 Formula was a real looker. At a cost of only $100 over that of an Esprit, the model was definitely worth the price. It was set off by the sweeping fiberglass hood with the familiar twin air scoops traversing its length. There was also suspension to handle the performance, with a brawny 1-1/8in front stabilizer bar, heavy-duty shocks, and the F70x14in rubber.

For many, the 1972 Firebirds were the beginning of the end of the late-1960s/early-1970s muscle era. Granted, the emphasis had moved toward economy and image during that year, with a turn away from performance. The performance, however, was still there (hidden in those net horsepower numbers), though it was ignored in company advertising.

A 1972 Firebird certainly shouldn't be overlooked if performance (still plenty there) and classic styling (still a lot there, too) are your goals.

1973 Firebird

The eight Firebird models for 1973 sported an abundance of new interior motifs, revised colors, and slightly modified sheet metal, but the big news for the year was the introduction of the killer SD-455 powerplant. This last of the muscle mills for the decade was available only in the Formula and the Trans Am.

The introduction of this powerplant (which would be available for two model years) was somewhat of a conflict of attitudes and atmospheres. First, by 1973, horsepower was no longer in vogue, so its presence wasn't even advertised. Racy looks were OK, but most engineering efforts were directed at reducing gas pump dependency. That's why the appearance of this powerplant was so interesting for the period.

But the Firebird went forward in this bold direction, practically standing alone, and certainly leaving behind its Camaro brethren, and Mustang competition, which had long since shed their big-

The look of the front end for the 1972 Fire-bird wasn't really that much different from the 1971 model. The big difference was in the honeycomb mesh used for the inserts. The macho hood scoops were still available in certain models.

blocks. Not only did PMD come out with the Super Duty version of its 455, but it also continued the standard 455 version.

Below the 455 engines, though, came a considerable drop in performance, following the direction of the rest of the industry with falling horsepower and compression ratios. The power fell below the 1972 performance levels, which also had shown decline.

Total Firebird production (including Trans Am) was a dismal 46,313 for the year, with the sporty Formula being the lowest of the three Firebird models. The Esprit sold the best at 17,249 units.

Safety was having a big impact at the designers' desks, resulting in a re-designed nose for the Firebird. It was still built of the Endura material, but had been re-engineered to be much more crash-worthy. All Firebirds carried this new design.

Again, the base Firebird could be ordered with either the Code D six-cylinder or Code M V-8. Interestingly, even with the consumer's concern for fuel economy, only 1,370 of the 14,096 base models sold carried the more economical six-cylinder engine.

Power was not aplenty in either engine. The 250ci six-banger provided only 100nhp and 175 lb-ft of torque. The bottom V-8 managed only 150nhp and was equipped with the first sub-eight compression ratio (only 7.6:1). Surprisingly, it produced a potent 270lb-ft of torque.

With the signature scripts on the roof pillars, the Esprit was readily identifiable this model year. There were also nice touches when you opened the door, like an African mahogany dash and accent

Next page
Following the top-of-the-line Trans Am came the Formula, keeping much of the glitz of the previous year. The model was now called just Formula, with the Firebird name nowhere to be found. It was an extremely low-production model with only 5,249 produced, and only 1,082 carried the manual transmission.

panels on the center console. Hidden wipers and color-coordinated mirrors gave the Esprit a racy look.

But it lacked the power of past years to match its appearance. The same baseline Code M 350 provided economical two-barrel power, but performance? Forget it!

The Code R 400ci mill was also available as an Esprit option, a combination called the Esprit 400. The extra cubic inches brought twenty more horsepower over the standard 350. With an 8:1 compression ratio, the two-barrel mill delivered an impressive 320lb-ft of torque. All Esprit 400s carried automatic transmissions. Although certainly not a muscle machine by 1990s collector standards, the Esprit 400 could prove to be an interesting acquisition, especially as the big-block, high-performance models become more rare and costly.

The 1973 Formula was a real bargain, as its base price exceeded the Esprit's by only twenty-seven dollars. That paltry sum bought a custom twin-scoop hood, a black textured grille, heavy-duty suspension, dual exhausts, and a custom steering wheel. At a base price of only $3,276, this was a real buy. Undoubtedly, these Formulas are going to increase dramatically in value, especially those powered with a 455.

Both the 350 and 400ci versions of the Formula (known as the Formula 350 and Formula 400) produced more grunt than their Esprit counterparts. The Code N 350 was up twenty-five horses to 175, while the 400's performance was 230 ponies. The extra performance came mostly from the dual exhausts and four-barrel carbs of these engines.

Then came the two versions of the 455ci powerplants. The Code Y version was rated at 250nhp, down from the previous year due to a reduction of compression ratio to 8:1.

The introduction of the new Firebird was announced by this national advertisement. This was the final year of this body style, which remains popular with collectors, especially those models powered by the awesome SD-455 engine.

Pontiac's '73 Firebirds are here.

ticed,

n.

onal. It's not

.!

ove. The

Order the

the handling

While the scoops look tough, the toughest part of any Firebird is the front bumper. It's made of Endura to help fight dents and dings. And it's been reinforced this year to make it stronger.

Esprit: Can a sports car be luxurious?

Esprit wipes out all doubt. The new bucket seats, the new cloth or all-vinyl upholstery, the new instrument panel and door trim are as plush as you'll find in many a luxury car.

The ride's almost that plush, too.

Basic Firebird: What we didn't sacrifice for price.

This is our easiest to own Firebird.

You still get molded foam bucket seats; loop-pile carpeting; High-Low ventilation; the Endura bumper; a strong, double-shell roof that absorbs sound; Firebird's futuristic styling and outstanding handling.

That's our way with sports cars. Are you ready to get serious?

GM
MARK OF EXCELLENCE
Pontiac Motor Division

Buckle up for safety.

The Wide-Track people have a way with cars.

The front end of the 1974 Firebird had a gentler look than in the previous year. The styling was slanted and carried a black

bumper guard. There was also a new treatment of the parking lights, with the "X" crossed chrome bars of the previous year

eliminated. The twin grille openings featured vertical bars, with the Pontiac lettering in the left opening.

Then came the big guy! It was surely a duck out of water in an era of decreasing performance. The super-powerful (and today super-desirable) Super Duty 455 Code X powerplant made performance fashionable again. The specially reinforced block featured forged rods, aluminum pistons, dry sump oiling system, special cam, four main bearings, and a dual exhaust system. The net horsepower was rated as 310, but if measured in brake horsepower, that figure could have been 350 or even greater—a number very comparable to the big-block maulers of the late 1960s.

Originally designed for the Trans Am, PMD decided to make the powerplant available for the Formula as well. The version was called the Formula SD-455. Only forty-three were constructed and very few have been located, making it the most desirable of the Formula

breed, bar none. Having this engine under the hood, with that "SD-455" lettering on the cowling, could nearly double the car's worth over that of the non-Super Duty Formulas.

Acquiring one of these SDs cost the buyer plenty at the time. It took an extra $675 (huge money in 1973) to get it under the hood.

But if performance was your bag, the SD could answer the call with high-13sec quarters at over 100mph.

1974 Firebird

For many Firebird fans, the design changes for the 1974 Firebird just didn't make it. It was as though the stylists' input had been totally ignored and the results from the wind tunnel were taken as gospel.

The sporty, flat-nose design, which had been with the model since its incep-

tion, was now replaced by a slanted design. The twin grille styling was still in place, but much of the pizzazz had departed.

There were also considerable changes made in the rear of the body. First, the chrome rear bumper was no more, replaced by a rubber impact strip that traversed the complete width of the rear and rolled around only to rear quarter on both sides. Also, the taillight treatment was altered, with the lighting stretching further inward toward the center and the backup lights bracketing the license plate area.

Without a doubt, the Firebird had a new look. Surprisingly, popularity was up from the previous year, with a total (including Trans Am) of 73,729 sold. The Esprit was by far the most popular model, selling 22,569. The base Firebird (with V-8 power) sold 18,769. The again-stylish Formula showed strong sales of

44

The powerful SD-455 engine was the top engine in 1974 for both the Firebird and Trans Am. This was really the last of the muscle powerplants in the 1970s, and it produced 290nhp—long believed to be underrated. This rare powerhouse, which produced 395lb-ft of torque, was carried in only fifty-eight Formulas, making those machines the best of the breed for 1974 collectibles.

14,519, but only fifty-eight of that number carried the powerful SD-455 mill.

The base model was again offered with either the in-line six or V-8 powerplants. The six-cylinder version sold 7,603. The Code D six-cylinder barely made it to three figures with its net horsepower rating sitting exactly at 100. The optional Code M 350ci V-8 carried a two-barrel carb, single exhaust, and a 7.6:1 compression ratio, which enabled 155nhp. Later versions of this powerplant reportedly made an additional 15hp. Torque was rated at 240lb-ft at only 2400rpm.

If the buyer supplied another $500 beyond the base model's cost, Pontiac

would add a number of trim items and provide the Esprit model. The sporty machine carried the expected pillar badge, custom steering wheel and interior package, concealed wipers, color-coordinated door handles, wheel and sill mouldings, and deluxe wheel covers. One classy little machine!

The Esprit might have looked fast just sitting there, but it was looks with no substance when the standard powerplant was fitted. The Esprit carried the same 350ci powerplant as the base model, but the 400ci/175nhp two-barrel version could also be acquired for an additional fifty dollars.

With the 400 engine, the Esprit was a nice appearance and performance combination for a very economical price tag. A total of 22,583 copies were sold, so there should still be a number of these cars around in the 1990s.

For 1974, the Formula and the Trans Am were once again the heavy hitters for the performance-minded. Both were promoted as "sports cars," but there was still muscle-type power aplenty under the hood.

The Super Duty 455 remained the blaster it had been the year before (with the appropriate SD-455 nomenclature on the hood scoop), but again, only a minuscule fifty-eight of these hustlers were produced. Rare? Desirable? Expensive to buy in the 1990s? The answer is "yes" to all those questions, as these models will continue to escalate in value.

One of the best SD-455 Formulas in the country is owned by Wolfgang Feist. Most of the time it's mistaken for a Trans Am, but the 1986–1988 National Champion machine is a Formula. This is one of that magic fifty-eight, and it's something to see.

"The Formula normally had the dual sloping hood design, but with the Super Duty option, it had the shaker hood configurational. The decal on the shaker was the only identifier of the SD engine, with the exception of an 'X' in the VIN number," Feist explained.

Classified officially as the LS-2 powerplant, the SD-455 powerplant carried the same 290nhp rating that it had in 1973. But with its high-performance innards, it was a performer of monumental

proportions. The quick-revving, 8.4:1 compression ratio engine brought frightening response when the pedal was slammed to the floor. The SD-455-powered Trans Ams seemed to get all the publicity, but this Formula deserves its place in the spotlight too as a candidate for being one of, or the most, desirable of the Firebird breed.

A lesser 455 powerplant was again available for the Formula that year. The L-75 model produced only forty fewer horses than its SD brother. No slouch, the L-75 pumped out 395lb-ft of torque, only fifteen fewer than the Super Duty.

An interesting footnote to the 455 story is that both versions of the 455 were dropped late in the model year. Quite simply, performance just wasn't selling. Although many felt this would be the big-block's final bow, the 455 remained in the line-up for two more years.

But if your manhood didn't require a 455, there were two other engines available for the Formula in upgraded versions of the 350 and 400ci engines. A 170nhp version of the 350 engine, along with a 400ci engine capable of 190 horses, provided the Formula purchaser with many choices for his performance whims.

As was the case with previous Formulas, the model carried a number of options to match its performance image. There were dual exhausts, a heavy-duty suspension, and wide tires. The double scoop grille was still in place (except in the SD versions) along with the distinctive, blacked-out grille.

It was the height of the gas crunch, a period that had killed most muscle cars, but the Firebird was still hanging in there with significant performance to match its muscular looks.

The Firebird design for 1974 was a buyer favorite and came out on top in a national poll. The rear-end treatment was completely restyled, with horizontal bars stretching almost entirely across the tail. The model looked completely different from the previous year—but it was still a Firebird.

1975 Firebird

Was reduced performance the direction to take the Firebird in 1975? That was the question that PMD had to answer, and the final decision was to continue the lower power trend.

There wasn't a single Firebird V-8 engine that had an 8:1 compression ratio in 1975. The lone 455 powerplant, which had returned to the line-up by popular demand, could no longer be ordered with the Formula and was restricted solely to the Trans Am. Even more unsettling, it was rated at only 200nhp, some ninety horses short of the Super Duty of the previous model year.

The Firebird models available (excepting the Trans Am) were halved from the previous year, leaving only four: the base model, one Esprit, and two Formulas. Despite fewer model choices, it was a very good year production-wise, with the six-cylinder base Bird being the top seller at 22,293, while the Esprit was close behind at 20,826. Both the base and Esprit models were available with either the Code D six-cylinder or the Code M V-8 (Code J in California). Formula sales lagged the base and Esprit Firebirds, finishing a distant third at 13,670. Looking at the sales, one might assume that higher performance was on the way out, but

that certainly wasn't the case, as the Trans Am led all the models with an impressive 27,274 total.

In an interesting contradiction of the downward performance spiral, the six-cylinder mill continued to show a slight increase in power. The compression ratio moved up a notch to 8.25:1, with a five horsepower increase to 105 net horses. The 350 engine was identical to the previous year's powerplant, but it showed an additional five pounds of torque at 280 lb-ft.

As for the 1975 design, there was little change from the previous year. The front end treatment was almost identical,

The macho look of the previous years seemed to be gone. This Formula carried the hood-scoop design that traversed right to the front of the hood.

48

the only difference being that the slats in the twin grille cutouts were now horizontal instead of vertical. There was also a new roofline design.

A pair of three-word systems was all new for 1975 and standard on all models: Radial Tuned Suspension (RTS) and High Efficiency Ignition (HEI).

The base Firebird was pretty much a "Plain Jane," but there were two interesting items of standard equipment: a common, two-spoke steering wheel and moon-style wheel covers.

More luxury and a little more power came with the Esprit. Depending on the engine selection, the base price could reach $4,088. The normally expected Esprit options were in place in the form of decorative mouldings, door handle inserts, concealed wipers, and pillar scripts.

The Formula line again carried the Formula 350 and Formula 400 nomenclature. The 350 powerplant, featuring a four-barrel in this application, was rated at 175 horses. The L78 400 engine would be the top mill for the Formula dropping in 1975 to 185nhp. This continuing downturn in power certainly had the performance-minded hanging their heads in sadness. But then again, it was happening throughout the industry. There was really a double-whammy to what performance there was, because the lower rearend ratios actually made the figures lower than they seemed.

Not only did the 1975 Formula not have the performance capabilities of the earlier models (in fact, not even close), but the general opinion was that the looks had also deteriorated. The twin scoops seemed to be out of place with the front end treatment of this model. Also gone was the swoopy, dashing look of the past. With the decreased performance, the Formula just wasn't the same anymore.

It was almost the end of the performance era and there were many who were sorry to see its demise. But that was

the trend in the mid-1970s, and it seemed that all manufacturers were going along with it in varying degrees.

1976 Firebird

America's bicentennial celebration year was also somewhat of a landmark year for the Firebird line. A number of significant changes took place, events which would change the model's direction in the years to come.

Nineteen seventy-six would be the first year that production would exceed six figures at 110,775. The breakdown showed an almost equal preference for the base model, Esprit, and Formula, but the real enthusiasm was for the Trans Am.

Perhaps the biggest transition was the fact that this was the final year for the infamous (although you could hardly describe this downgraded version in that manner) 455 big-block. It was already gone in the Formula, but its final breath came as a performance option with the Trans Am.

There were also considerable styling changes (highly acclaimed by the car magazines at the time) which gave the Bird a quite different look. The lower spoiler inlets were decreased in size and now contained the parking lights, which had earlier been located in the upper grille openings. Another real double-take was the bumper treatment, which now carried the body colors and gave the machine a more integrated and sweeping look.

The base model, which provided sporty looks and good economy, could be acquired with the standard L-22 250ci L6 powerplant whose horses had been kicked up a notch to 110nhp, with a slight increase in compression ratio to 8.3:1.

Three additional mills were also available with the base model, which was certainly a change from earlier years. First, there was a pair of 350ci engines:

the L-30 two-barrel and the L-76 four-barrel version. The powerplants were worth 160 and 175 net horsepower, respectively, with the more powerful L-76 adding $195 to the price.

Surprisingly, one could also acquire the 185nhp L-78 four-barrel carb version of the 400ci powerplant, making the bottom-of-the-line Bird one very respectable performer. Clearly the new 400 was not on par with 400s of the past, but it was no slouch either.

For the first time, the Esprit options were identical to those of the base model. Now they just rode in more plush surroundings. The Esprit was a spiffy-looking machine, its Esprit logo carried on the rear of the front quarters and Firebird emblems emblazoned on the vertical roof stanchions. A flashy trim package swept from the upper front fenders across the top of the doors to just under the glass, and then up over the roof. Nice touches, indeed!

The Formula sustained major changes, and with them, seemed to have taken on more of a European sports look. First, there was the stylish (though nonfunctional), new, twin-scoop hood. Other options included side-splitter tail pipes and a new, full-length console.

But if you really wanted your Formula to dazzle the crowd, then you had to have the Formula appearance package. The "W50" option included a horizontal, blacked-out band along the bottom of the body between the wheelwell cutouts with large block-letter Formula lettering within this area. The air scoops got the same blacked-out treatment. Four different color combinations were available with this option.

Engine choices for the Formula were exactly the same as with the two lower models and certainly weren't up to the racy looks of the machine. The two-barrel 350 was the standard engine, with the L-76 350 and L-78 400 also being available.

Trans Am 1969–1976

1969-1/2 Trans Am

It was a Firebird, true, but the family resemblance was about all the Trans Am shared with it lesser siblings.

Chevy had the Z28 Camaro as its super pony car sports model, and a competitor with a PMD label was definitely needed. John DeLorean, Pontiac's top man at the time, appointed talented young designer Herb Adams to address the situation (actually Adams had volunteered for the job).

Instead of a completely new design, the decision was made to accomplish significant improvements to the existing Firebird. The project name was "Pontiac Firebird Sprint Turismo" (PFST). Vast R&D went into the prototype PFST, with concentration on suspension and steering refinements. A number of different powerplants were also examined.

The Trans Am name came from the popular, factory model-oriented road race series of the same name. PMD had to pay the series' sponsors, the Sports Car Club of America (SCCA), five dollars for every Trans Am sold. Use of the name,

however, did not meet with universal approval from the fans of the racing series despite the fact that Trans Ams would later become participants in the racing.

Many fans felt the name should be used only by the series.

The Trans Am had its original introduction at the Chicago Auto Show in

The Super Duty was the final grasp for power in the 1970s as fuel economy and pollution laws were strangling the breath from the muscle machines. This was really the last of muscle-car era powerplants.

This particular 1969 Trans Am carried the 400ci Ram Air III powerplant, as can be ascertained from the "Ram Air" lettering on the hood. The initial Trans Am wore the overbody striping, rear deck spoiler, hood scoops, and front quarter air extractor scoops. Steve Hamilton of Indiana owns this magnificent example of the first Trans Am.

There was almost total agreement from enthusiasts and magazines concerning the super looks and style of the new Trans Am rearend treatment. On this example, the stripes sweep over both the rear deck and the top of the rear spoiler.

March 1969, therefore relegating the Trans Am to status as a 1969-1/2 model. Only 697 built that model year, making them very rare and valuable in today's market. Of that small first year production run, a mere eight were convertibles.

Like other super-desirable vintage muscle machines, there suddenly seem to be a lot more than eight Trans Am convertibles around. Let the buyer beware, because there are obviously some phony 1969-1/2 ragtops floating around.

The cost for that first Trans Am was a significant $724 (the WS4 Option) over the base model. But if you just happened to have retained that purchase to this day, it was extra money well spent.

A sidebar to the Trans Am story is that the first TA was also marketed during part of the normal 1970 model year (the final months of 1969). In fact, it is reported that some of these cars might have been titled as 1970 models.

Appearance-wise, the Trans Ams all came in Cameo White, with blue striping that swept from the hood scoops to the

**Announcing
Pontiac's new pony express.   Firebird Trans Am.**

Back when the Chisholm Trail was considered an expressway, you needed 335 horses to haul the mail. We figure you still do. So Firebird Trans Am's got 'em. Stabled under oversized hood scoops in 400 cubic inches of Ram Air V-8. A heavy-duty, 3-speed box hitches them to a 3.55:1 rear axle and fiber-glass-belted tires. Wells Fargo rides again!

Sound like Trans Am is strictly for wide-open spaces? Take it through a mountain pass. Heavy-duty shocks and springs, 1" stabilizer bar, power front disc brakes and variable-ratio power steering make Trans Am our version of a quarter horse.

But you can probably guess all that by looking at it. Trans Am's engine-air exhaust louvers, rear-deck airfoil, black textured grille, full-length blue stripes, leather-covered steering wheel and special I.D. provide fair warning that this is no ordinary mount. It's Pontiac's new pony express. And that's about as far from ordinary as you can get.

Shown above are some of the many available Trans Am features. See your Pontiac dealer. Pontiac Motor Division.

The new Trans Am hit the country to great national fanfare. The model's race car styling and dashing striping really set it apart from the crowd. This early national advertisement called it "Pontiac's new pony express." That it was.

This Firebird symbol was carried over the rear deck keyhole on the initial Trans Am. It was omitted the following year as the Trans Am model moved away from its Firebird roots.

back of the rear deck. On some models, the striping ran over the deck-mounted rear wing, although company literature said otherwise.

That sporty rear spoiler was there for more than just looks. Pontiac engineers reported that the airfoil pushed down with 100lb of stabilizing downforce at 100mph.

Also of great interest were the front, quarter-mounted (and functional) air extractor scoops, which looked like something right out of a Buck Rogers cartoon. The purpose of the scoops was to remove heated air from the engine compartment.

The Trans Am suspension was a beat apart from that of its Firebird brothers and featured staggered rear stocks, heavy-duty springs, 1in front stabilizer bar, super-quick steering, and 7in wheels. There was also a special steel hood carrying two functional, front-indented scoops. A foam seal on the underside of the hood made an airtight connection to the powerplant. The air from the topside scoops was funnelled directly to the air cleaner and then into the carburetor.

The rear end of the first Trans Am had the look of a real race car with its twin racing stripes run over the length of the body. There were actually two variations of the stripes: most Trans Ams had them running under the rear spoiler, but a few had them applied directly on the spoiler's top surface.

Two so-called "Ram Air" power-plants were available with the first Trans Am. The 400HO (Ram Air III) L-74 (with steel heads) produced an advertised 335hp. The optional aluminum head 400 Ram Air IV (L-67) provided an additional ten ponies. But that rating was an absolute joke, as the same powerplant in the GTO carried a rating twenty-five horses higher.

Of the 697 Trans Ams produced, 634 were outfitted with the Ram Air III option: 127 with manual transmissions, and 524 with automatics. Needless to say, the collector interest in the remaining fifty-five (forty-six manual, nine automatics)

On some of the initial 1969 Trans Ams, the Firebird name would be doubled with the Trans Am nametag. This was done only early on, as the company sought to give the Trans Am an identity of its own. The two names appeared on the model's front quarters.

The top powerplant was the Ram Air IV engine, followed by this Ram Air III mill. Also known as the 400HO, this high-torque pro-ducer pounded out an impressive 335hp, only 10hp fewer than the Ram Air IV engine.

RAM AIR IV

that had the L-67 Ram Air IV engine is very great. L-67-equipped Trans Ams are one of the most desirable muscle cars in today's market. It's certainly easy to understand how John Gunnell, in the *Illustrated Firebird Buyer's Guide*, awarded it a five star investment rating!

1970-1/2 Trans Am

There were a number of significant sheet metal changes made to the top Firebird for 1970. Probably the most noticeable, though, was that you could buy the racy machine in another base color besides white. Just flip-flop the blue and white and the new color scheme came to light: all-blue, with white stripes. Performance fans liked what they saw, because more than four times the number of 1969-1/2 Trans Ams (3,196) were constructed.

A full-front air dam (reportedly producing up to 50lb of downforce at speed) was attached below the grille and flared

up over the front of the wheelwells. It gave the machine a look of, well, a "Trans-Am racer." The wheel wells were flared race car-style, as well. Also, an integrated rear spoiler blended magnificently with the design's overall appearance.

Great looks, you bet, but plenty of aerodynamic research had gone into the shape of those muscular-looking protrusions. The wind played sweetly over them, making this second Trans Am one of the sleekest machines produced for its time period.

The characteristic Trans Am block lettering was inscribed for the first time on the rear deck, just below the spoiler. It was a position the insignia would occupy for many years to follow.

The motoring press reports on the model all acclaimed greatness. *Car And Driver* described it as "A hard muscled, lightning-reflexed commando of a car, the likes of which doesn't exist anywhere

in the world, even at twice the price." Some pretty heavy words, but they were certainly deserved by this killer machine.

As in the previous year, the late introduction of the model (March 1970) again added the "1/2" to the model dating. As in the initial year, the model featured the two top 400ci powerplants: the Ram Air III and the Ram Air IV.

The Ram Air III was the standard mill and provided 335bhp at 10.5:1 compression ratio, only five horses more than the Formula's baseline powerplant. But even though it wasn't the top powerplant that year, there was still amazing performance available, as a February 1970 *Hot Rod Magazine* test revealed.

The Trans Am, which was equipped with a four-speed and 3.90 gearing, made an unheard of 13.9sec run at 102mph. To prove it wasn't a fluke, the car was raced down the strip a total of four times, for a 14sec average clocking. *Hot Rod* noted that the best quarter mile performance

The 1970 Trans Am looked like a full-race machine with its integrated front air dam and side fender flares. The bird was carried on the top of the body-colored grille, with a new striping treatment. Rally II wheels were standard equipment on this model, as was a redesign of the air extractor scoops.

was attained when the final shift was made at 5500rpm, "even though the engine is good for 6000."

But for those who wanted that "little extra," there was the special-order Ram Air IV. Only eighty-eight of these were purchased. The IV provided an additional ten horses over the Ram III engine, and it goes without saying that to muscle enthusiasts, then and now, this is a very, very desirable vehicle. Both Ram Airs used the Rochester four-barrel Quadrajet carb. Torque was rated at 430 lb-ft for both powerplants, but the figure came at 300rpm higher for the IV version.

Both Ram Airs looked as good as they ran, with chrome valve covers and the characteristic shaker hood that was quite the rage with the performance set during this period.

Besides its most obvious characteristics, there were also numerous other nice styling touches that set the Trans Am apart. For example, the front-quarter air extractor was completely redesigned and relocated higher on the front fender. The "Trans Am" lettering was moved from its

former forward position to just behind and below the air extractor and the fender crease.

The redesigned interior was up to snuff with the outside. The classy aluminum instrument panel made sweeping display in front of the driver, with the tach rotated in such a way that the redline zone would be at the top of the dial.

Other Trans Am standard equipment included concealed wipers, dual horns, dual outside mirrors, Rally II wheels, and power brakes and steering.

The superb handling of these magnificent machines was the result of the Y-96 suspension system. A monstrous 1.25in front sway bar held the front end in check, assisted by a 0.875in sway bar out back. The package also included heavy-duty springs and shocks.

For many, things just couldn't get any better than the 1970-1/2 Trans Am!

1971 Trans Am

Nineteen seventy-one was a year of transition for the top Firebird, with engine compression ratios starting a downward spiral and taking horsepower levels down with them. But there would be a change to the powerplants to compensate somewhat for that downgrading.

What was decided by PMD was to kick up the cubic inches to 455, an unheard-of figure for a pony car. Surprisingly, it was the only displacement powerplant fitted to the Trans Am.

With an 8.4:1 compression ratio and the functional, cool-air induction system, the LS-5 455HO powerplant provided 355bhp at 3500rpm. An impressive 480lb-ft of torque was also available. A 455HO decal on the shaker scoop identified the engine.

The Trans Am's production that model year, though, dwindled to only 2,116 built: 1,231 automatics and 885 manually shifted.

There wasn't much done externally to separate the Trans Am model from the

The 1971 Trans Am front-end design featured the wide hood stripe that traversed the center of the hood, including the Shaker Hood. A small version of the Firebird decal was centered on the nose.

The standard powerplant for the 1971 Trans Am was the punchy 455 High Output engine, denoted as the "455HO." It carried advertised performance of 335bhp and 480lb-ft of torque. The engine used 8.4:1 compression cylinder heads and benefitted from a cold-air induction system. A "455 H.O." decal on the Shaker Hood identified the powerplant.

The standard wheel for the Trans Am in 1971 was this stylish five-spoke configuration. The red center dome carried the PMD identification. In addition, a new honeycomb wheel was available for the first time with the 1971 model.

The interior of the 1971 Trans Am was updated from the previous year, and was certainly dashing and sporty, to say the least. The horizontally ribbed silver dash carried a montage of instrumentation, while the high-back bucket seats were among the best in the industry. The center console was standard equipment.

Changes for the 1971 Trans Am were minimal, but with the looks this stylish front end carried, why would you want to change it? The flared-in front spoiler incorporated air openings, with the parking lights located directly below the single headlights.

previous year's. A slight change to the Rally II wheel design, and high-back, more modern-appearing bucket seats, were about it. You also had to pay extra for the chrome engine dress-up parts if you wanted any glitz under the hood.

Although these cars probably don't quite match the appeal of some of the earlier Ram Air machines, the 1971 could well be an extremely valuable muscle car. Even though it is considered a transitional model marking the turn away from performance, the car's big-block engine and its classic styling make it a solid collectible sure to grow in value in the coming years.

Bob Pollock, of Columbus, Ohio, a muscle car financing expert, owns one of these magnificent machines. "I think that this model will continue to grow in value in the years to come. I'm sure keeping mine because of that belief," he said.

1972 Trans Am

Significant labor problems at the Norwood, Ohio, assembly plant wreaked havoc with Trans Am production for the 1972 model year. With only 1,286 of the model produced, it was the smallest run for the Trans Am (with the exception of the initial 1969 model year). The low production number in concert with twen-

ty-plus years of neglect, abuse, crashes, and other unfortunate events has made these models very difficult to locate.

One of the strange sidebars to Pontiac's labor problems was that they occurred at the end of the model year and left a number of partially completed Trans Ams (and Camaros) stuck on the assembly line. Since these cars would have had to be considered as 1973s if they were completed, they were either destroyed or donated to schools because the new 1973 pollution standards could not be met. Such a waste!

There were practically no changes made to the 1972 Trans Am. By hanging

The striping for the 1971 Trans Am covered the rear deck and ran up the hoodside of the integral spoiler. The "Trans Am" lettering was carried on the rear of the spoiler. To many, it was one of the best treatments of any of the Trans Ams.

It was easy to discern the existence of the High Output 455ci mill in the 1971 Trans Am. With a net horsepower of 305, the per- *formance was still under the hood, but the decline would soon begin.*

onto its performance image even after the muscle began to atrophy in other muscle cars, the Trans Am was already different from much of the competition.

Again, there was only a single powerplant available: the Code-X 455HO. On paper, it didn't appear to be quite the performer of a year earlier, with a rating of 300hp and 415lb-ft of torque. But remember, those figures were net values,

Previous page
Not only do the 1972 Trans Am's dynamite looks set it off, but its rarity makes it highly collectible. In terms of appearance, the 1972 model wasn't that much different from the previous year's. The only powerplant available was the impressive 455HO mill, which provided 300nhp and 415lb-ft of torque.

making them roughly equivalent to the values of the previous year.

Road tests showed the model to be a solid 14sec performer at almost 100mph. With only slight modifications, 13sec runs were easily attainable. These were magnificent performance numbers, considering that the Trans Am weighed a beefy 3,564lb.

Both a Hurst-controlled four-speed and the M-40 Turbo 400 automatic were available. The Turbo 400 was the overwhelming choice, with 828 sold, compared to only 458 with the synchromesh.

A vast array of equipment came standard on the Trans Am, but an even larger quantity of options was available. The 1972 Trans Am had the same standard features as the Formula, plus the Formula steering wheel, a custom aluminum dash,

front air dam, wheel flares, rear spoiler, shaker hood with cold air induction system, Rally II wheels, front disc brakes, power-flex cooling fan, and the Safety-T-Track rearend. The standard suspension system again featured a 1.25in front stabilizer bar and a 7/8in diameter rear unit.

The 1972 Trans Am still featured swoopy looks and performance, but its numbers were few. One of these low-production TAs could make a very interesting collectible.

1973 Trans Am

For 1973, the magic words for the Trans Am were "Super Duty," which signified that performance was once again alive and well in the PMD camp.

The power rating numbers of that awesome LS-2 engine rested at an im-

Trans Ams for 1971 were available in the familiar Cameo White with blue striping, but also with a reverse blue body color with white striping. Changes to the body styling were minimal for the third Trans Am, but it still carried classic looks. This car is owned by Dwight Stump of Ohio.

pressive 310nhp, putting it back in the ballpark of the earlier Ram Air mills. But alas, the output would drop by twenty, owing to a failure to meet the emissions requirements for the model year. The power reduction was accomplished by substitution of a different cam.

The amazing aspect of the Super Duty concept was that the engine was making those good horsepower numbers at only an 8.4:1 compression ratio. It was truly an impressive engineering accomplishment.

Even with the dictated horsepower reduction, the powerplant (which was also available in the Formula) was a high-revving performer. The engine sported four-bolt mains, a stronger block with provisions for dry sump oiling, a Quadra-jet carb, forged rods and pistons, and high-flow (revised Ram Air IV) cylinder heads.

There was also a new version of the Ram Air IV cam with 308/320 timing.

This factory photo demonstrates the details of the practically unchanged 1972 Trans Am. The model featured the complete over-body stripe, which included the Shaker Hood, front quarter air extractors, and the stylish cast-aluminum honeycomb wheels. The wheels came on the scene for all models in 1972. Pontiac Motor Division

Styling was the highlight of the 1973 Trans Am with the air extractors and fender flares. The 455-SD nomenclature was carried on the Shaker Hood. To many, the combination of styling and the Super Duty powerplant made the 1973 Trans Am the best there ever was in muscle cars. Its value today certainly proves that theory. Bill Streeter

The SD-455 powerplant was around in 1974 for the last time, still producing 290nhp and 395lb-ft of torque. The so-called "LS-2 Option," which included this powerful mill, was on its way out. It's extremely rare, with only 943 so-equipped Trans Ams produced in 1974. Should you find one of these machines, grab it up.

The NHRA factored a higher horsepower rating for the Super Duty for its competition. It was just like the good old days!

Besides the ground-pounding powerplant, the Super Duty package for the Trans Am consisted of a beefed-up, three-speed Hydra-matic with a shift point at 5400rpm; a heavy-duty radiator; and a special suspension system featuring a front sway bar and heavy-duty underpinnings.

The performance was breathtaking. Bob Blair, of Vandalia, Ohio, bought one of the SD-455 Trans Ams brand new and pushed it through the quarter at an eye-popping 13.90sec at 106mph. "And that was done on street tires," he explained. "It was the gas crunch time period and I was told at the time that the model was not available. Even though I worked for General Motors at the time, I had to beg for it. Ordered it in February, but didn't get it until July." Easy to understand, as there were only 252 Trans Ams equipped with the SD option, 180 of which had the automatic transmission.

The standard Code-Y L-75 455 pow-

The bird decal for the 1973 Trans Am almost completely encompassed the hood area. The graphic really helped set off the Trans Am as a unique machine. For 1973, the Trans Am could be ordered in three different colors. Besides the Buccaneer Red shown here, it was also possible to purchase a Trans Am in Brewster Green or Cameo White. The styling of the 1973 TA was classic, making it one of the most popular of the vintage muscle machines. Bill Streeter

erplant was fitted to 4,550 Trans Ams. The engine carried a four-barrel carb and dual exhausts, and produced a 250nhp rating.

Style-wise, 1973 was also the first year of the "Big Bird" that would almost completely cover the hood's surface. Initially, there was some PMD corporate resistance to the design, but then it was accepted by the masses and would remain a Trans Am mainstay until the 1978 model.

For the more inhibited, though, there was a much smaller version of the Trans Am insignia adorning only the nose of the car. But the design itself was identical to that of its larger brother.

The Trans Am also carried the expected multitude of options, including a Rally gauge cluster, Formula steering wheel, front disc brakes, power steering, front wheel flares, front-quarter air extractors, and an integral rear spoiler.

Nineteen seventy-three was also the initial year that the fabulous Trans Am could be ordered in a number of new colors. Besides the traditional Cameo White, there was now a Brewster Green and the very popular Buccaneer Red.

Also, the racing stripes were discontinued. The Trans Am nomenclature was blocked on the front quarter just below the air extractor on each side, and even more prominently just below the rear deck spoiler.

If you see a 1973 Trans Am Super Duty for sale, snap it up; it could well be the hottest collectible from Pontiac for the 1990s.

1974 Trans Am

The "look" was gone. That vertical slated grille that seemed so much a part of the overall Trans Am styling had departed. It was as though the wind tunnel

The SD-455 was the top engine for the Trans Am for 1973. The front end again featured the sporty spoiler beneath the grille. The high-performance Super Duty powerplant was the hottest powerplant available that model year. This particular SD-455 has been restored without the bird decal normally carried on the nose.

data results had spoken, and the design instructions were to "slope the front end for better streamlining."

There was also a major rearend change. For 1974, there would no longer be any chrome bumper out back. Instead, it was replaced by the now-standard rubber bumper. Most enthusiasts found the rearend change a lot easier to live with than what PMD did with the front end.

The extensive Trans Am package was again in place, offering everything but the kitchen sink. Included were power steering; front discs; limited slip rearend; chrome extended dual exhausts; Rally gauges, including a tach and clock; integral full-length rear deck spoiler; air deflectors; front fender air extractors; Formula sport steering wheel; Rally II wheels; sport suspension; outside mirrors; and the 225 horse version of the 400ci engine. The base price was $4,350, but if you wanted more under the hood to go with that macho appearance, then you had to put down the extra bucks for one of the two available 455 powerplants.

The Super Duty 455 engine was available for the second and final time, and the powerful mill was highly advertised. But its sales were low for the Trans Am (943, to be exact). A manual transmission was fitted to 212 of the SD-455 equipped Trans Ams.

The SD-455 was easily the highest performance engine in the market. But the time was no longer right for such outrageous horsepower numbers. With gas being expensive and scarce, it was almost sacrilegious in many people's minds to put the pedal down on such a powerplant. To that end, production of the mighty engine was ceased before the end of the model year.

To many, the deletion of those famous "SD" letters from the options list spelled the end of the muscle era in America. The Super Duty—with its four-

1974 Pontiac Firebirds.

Part engineering.
Part soul.

There's something about a 1974 Firebird you won't find in any fact sheet or spec book.

Because it's something you can't weigh or measure or touch. It's something you have to feel.

We call it soul.

Take the '74 Formula Firebird for example. Any spec sheet will tell you it comes with a 350, 400 or 455 V-8. A floor-shifted 3-speed trans. Performance dual exhausts. Hood scoops. Front disc brakes. And front and rear stabilizers. A very impressive list of features.

But no feature list can explain what it's like to drive a Formula 'Bird. Gauges set so you can read them at a glance. Controls positioned so they seem like they're extensions of your arms and legs. Response so quick it almost anticipates your commands. And an overall driving experience that makes it hard to suppress a toothy grin.

That's the soul of a Firebird.

The '74 Firebird Trans Am is even better. Because what we know about performance driving, we make standard on Trans Am.

A 400 4-bbl. V-8. 4-speed trans. Power front disc brakes. A limited-slip axle. Full instrumentation. F60—15 tires on 7" Rally II wheels. A shaker hood. A complete entourage of functional air dams, extractors, deflectors and spoilers.

And enough soul to make Trans Am the ultimate Firebird.

Anybody can appreciate Firebird's engineering. To appreciate the soul, you have to love driving cars as much as we love building them.

PONTIAC GM
Pontiac Motor Division

The Wide-Track people have a way with cars.

"Part engineering, part soul" was the way Pontiac described the 1974 Firebird in this advertisement. The company flaunted the Firebird's selection of powerplants, new suspension system, and revised body styling.

The 1974 Trans Am was basically unchanged from the previous year and again sported the popular hood bird. The rear of the 1974 Trans Am was revised, however. The chrome bumper had departed, to be replaced by the now-standard rubber unit. It all blended together beautifully with the hood spoiler and triple-slotted taillights.

70

The Super Duty powerplant was gone for 1975, but a new, so-called "455HO" engine made its debut. The new engine certainly didn't have the punch of the earlier engine of the same name: it produced only 200nhp. Externally, there were few changes made to the 1975 Trans Am with only minor alterations to the grille and rearend.

bolt mains, forged pistons, and special heads—was a wonderful swan song to the high-performance age.

It goes without saying that acquiring a Trans Am with that rare powerplant in today's market is a find of monumental proportions. The lovers of the vintage muscle sport will quickly tell you that both this SD and the 1973 SD Trans Am rate at the very top of the collecting desirability list.

The vast majority of the 1974 Trans Am buyers, however, found the more conservative 400 and 455 mills to their liking, with 4,664 and 4,648 sold, respectively. Although certainly not as desirable as the SD, these cars, especially those with the 455 version, also find favor with modern collectors. Both of these engine choices sported 8:1 compression ratios and opening model year horsepower ratings of 175 and 215 net horses, respectively.

1975 Trans Am

The Super Duty was gone for good and it seemed performance had become a four-letter word, but as far as Trans Am sales were concerned, it was probably just as well.

The conservative 455/200nhp version was the best you could get, and only 857

71

For a time during the 1975 model year, a 400ci mill was the biggest and most powerful motor you could buy for your Trans Am.

Later in the year, however, there would be a call for a 455, and PMD would respond with the standard, big sedan powerplant.

Trans Ams so-equipped were sold. The 400ci powered models showed an impressive 26,417 units sold. It seemed the message of economy over performance had been taken to heart by the buying public. But even though the Trans Am high-performance image was gone, the TA was still a popular seller (27,274), accounting for about a third of the total Firebird sales.

Actually, the model year started with the 400 being the biggest power under the hood. But there was still enough perceived interest to reinstate the Code-Y 455 mill, though it could hardly be compared with its earlier brothers. This 455 had only a single exhaust, and even a catalytic converter. This model year, the Trans Am was the only platform on which the downgraded 455 could be ordered. It was a pittance to be sure, but a little performance was better than nothing for those who craved it.

By the way, PMD actually called the big-block the 455HO. Really, though, the

mill was nothing more than the standard big-block engine that was used in the division's large sedans and station wagons.

But when you matched the powerplant against what Ford and Chrysler were offering at the time, the 455HO could definitely hold its own. Ultimately it comes down to what you're using as a reference point at the time.

Appearance-wise, there were very few changes made to the Trans Am from the previous year. Wraparound rear glass and slight changes in the grille were about it. But to many, the new grille design certainly didn't have the pizzazz of earlier versions and wasn't nearly as pop-

The displacement number was still the impressive 455ci, but the power was a far cry from the horsepower numbers of earlier years. There was only a single exhaust for the 1975 version, with a catalytic converter to cut into the power numbers of this once mighty mill.

Probably the most desirable 1976 Trans Am was a Limited Edition model released for Pontiac's fiftieth anniversary. The model was a real looker with its gold and black color scheme. The gold detailing was about as sharp as you could get. Only 1,947 units of this desirable model were produced, and they are a valuable collectible. The identification of the Fiftieth Anniversary model was clearly identified by the medallion on the car's front quarters. The Trans Am lettering was also done in a unique style for this particular model.

ular as the earlier model. The design could best be described as "refined" more than "rakish."

1976 Trans Am

In 1976, the Trans Am topped the

74

stylish Corvette in sales for the first time and to many became the number one domestic sports machine in America. The Bicentennial model was described in print as the top performance car of the year.

There were a number of firsts and lasts for the Trans Am during this model year. The most significant "last" would be the 455ci powerplant, which would make its final bow on the options list. As in 1975, the 455 was available only in the Trans Am.

Again, the 455 was a low-power version of the once-powerful mill, with a 200nhp rating and a lowly 7.6:1 compression ratio. The engine appeared in three different versions that year, with 7,099 installed in the base model and 429 residing in a pair of limited edition versions.

But even with the lower output big-block, the Trans Am was still an impressive performer, great for the time period and surprisingly not all that shabby when compared with the earlier muscle machines. The April 1976 *Car And Driver* reported a quarter-mile performance of 15.6sec at 90.3mph.

The 1976 TA carried all the performance add-ons, including front and rear deck spoilers, distinctive shaker hood, front-quarter air extractors, and Rally gauge cluster. The model also featured the popular twin exhaust extensions protruding together behind the left rear tire.

For the collector, there were two special edition Trans Ams (a coupe and a T-Top version) that will undoubtedly increase in value. The models were a sinister black, with a montage of gold detailing including gold honeycomb wheels, gold interior trim, and a gold body striping. They were lookers of the first order.

Pontiac saw fit to use the magnificent model as the vehicle to celebrate its fiftieth anniversary. The stylish machine was a great choice for the honor.

The Limited Editions were available with both the 400 and 455 powerplants. The T-Top models were coded Y82, while the coupes carried the Y81 nomenclature. The coupe equipped with a 400 engine was by far the most popular, with 1,628 constructed. Only 319 coupes carried the 455. Five hundred thirty-three of the lower-production T-Tops came with the 400; 110 with the 455.

The Limited Edition models weren't cheap by any means. The Y82 added a healthy $1,183 to the price tag, while the Y81 coupe version was about half that figure, at $556.

The Trans Am production total was 46,701, more than twice that of any of the other Firebird models.

It cost you an extra $103 to have a set of these honeycomb wheels, but to those who made the purchase, they were definitely worth it. PMD thought enough of the design to include it as the wheel cover for the Fiftieth Anniversary Trans Am Limited Edition model.

Firebird and Trans Am 1977–1994

It's hard to draw a line and call everything on one side of it a "modern" Firebird or Trans Am. But by applying an engine-based philosophy, the new realm could be considered to start with the 1977 models. The rationale behind this decision is that the 1976 model was the final Firebird to carry the 455ci big-block.

Goodbye 455

The top displacement for 1977 was the 185nhp 403ci L-80 four-barrel mill. The W72 version of the 400ci engine was rated at the 200 horse mark.

For 1977, there was also an extensive change in the grille: still the characteristic twin scoops, but the indentations weren't as deep as in the past. There were also four rectangular headlights for the first time. In addition, the Trans Am showed a broader bulge in a new hood design.

The 1994 lineup continued the 1993 styling redesign with two Firebird and two Trans Am models. The Firebird Formula was powered by the 275nhp 350ci LT1 whereas the Coupe came with the 160nhp 207ci L32 V-6. The Trans Am was fitted with a six-speed manual transmission and the LT1; the Trans Am GT shared the LT1 but was packed with luxury and performance options. Pontiac Motor Division

For performance fans, 1977 was not a happy year: the vaunted 455ci mill was no more. But the macho exterior design was still in place, even though there were some styling changes. The front-end changes for the 1977 Firebird line were sizable. First, there was the adoption of four headlights, then the addition of a new pointed nose and the Endura bumper panel. Pontiac Motor Division

Two big-block powerplants were available with the model in 1977, with both a 403ci and the W72 version of the 400ci engine. Power from the two engines was 185nhp and 200nhp, respectively. The days of the big horsepower engine of earlier years were definitely in the past.

There was an entirely new look for the 1977 Trans Am with a completely new grille design. The two grille openings were retained, but the inner portions of the openings were curved, providing a completely different appearance. There were also twin rectangular headlights for the first time. The model looked different, that's for sure. For the entire Firebird line, it was the third-bestselling year ever, with over 155,000 units sold.

This Formula shows its colors to excellent advantage. The Formula lettering was carried in the trim stripe on the lower portion of the body. For many enthusiasts, its looks were equal to those of the top gun Trans Am.

As in years past, there were special packages available for both the Formula and Trans Am. For the Formula, it was the W50 option that featured the lower body stripe, blacked out hood scoops, and the large "Formula" lettering. The Y81 Special Edition Package for the TA cost an extra $556, and with the Y82 option (a T-top configuration) there was a monstrous $1,143 extra cash required.

Nineteen seventy-seven was the third-best year for sales in Firebird history with a total of 155,736 sold (68,745 of which were Trans Ams) surpassed only by the two following years' totals.

Little was changed for 1978, with much of the 1977 design carried over. Appearance-wise, there was a new blacked out grille in the Formula which really set it apart. Only minor changes were made to the Trans Am, and most of these were made to the interior.

A new 305ci small-block was available for the first time, now a Chevy design. It was rated at 145nhp. There were also 400 and 403ci engines available.

Special editions abounded this model year, with the Esprit, Formula, and Trans Am being so-endowed. For the Esprit, there were the Sky Bird and Red Bird editions. Featured were special "Bird" decal identification, custom paint, and custom interiors. The W50 appearance package was again available with the Formula. But there were three special Trans Am options: the Y82, Y84, and Y88. None were cheap, all costing over $1,200.

The Y82 and Y84 Editions were similar packages that featured black and gold styling and offered either the 400 or 403 powerplants. The midyear Y88 Special Edition, though, has to be considered the most collectible Trans Am of the year. Coined the "Gold Special Edition," the version carried a Solar Gold finish through to the wheels. There was also unique body striping and a new Bird resting on the hood.

The Formula was one sharp package for 1979. The model was accentuated by the lower body black strip and the large "Formula" lettering contained within. The Formula also carried a heavy-duty suspension and blacked-out rear taillights. The Formula lettering also rode across the rear deck, and really set off this neat machine.

Another Facelift

For 1979, the grille tradition was finally discontinued. The characteristic split-screen design that PMD had revolutionized had succumbed to more modern styling. The new design featured twin headlights separated by vertical slats, with the characteristic center hood section retained. The change in design, however, didn't seem to bother the buying public because 1979 was the Firebird's greatest sales year ever at 211,454!

There was again a wide choice of powerplants, including both a two-barrel and, for the first time, a four-barrel version of the V-6. But the 400 and 403 V-8s still garnered the most attention and tallied the biggest sales.

It was also the tenth anniversary year for the Trans Am, and PMD made the most of it selling a best-ever 117,109. The Tenth Anniversary Trans Am package had both performance and appearance going for it. Under the hood of most of the models was the 403ci mill, this version producing an impressive 220nhp. The performance era wasn't quite over with those kind of numbers on tap.

But it was the looks of this machine that really set apart the first decade celebration. In the 1990s, this is going to be a very collectible machine. There was a

bigger bird on the hood, custom silver leather bucket seats, silver-tinted hatch roof, a special front air dam, and custom wheel covers.

The 1979 Trans Am featured a new front-end treatment, highlighted by four individual headlamps each set in their own enclosure. The lower portion of the grille also showed a change, with larger, twin, blacked-out openings. This Trans Am also carried the optional big bird decal.

The Tenth Anniversary model was also celebrated as a NASCAR Pace Car. This particular machine was used by Pontiac Division honcho Bob Stempel. Only a few of these models were built, and one was used to pace the 1979 Daytona 500. A highlight of the model was the Turbo aluminum-alloy wheel manufactured by the Appliance Wheel Company. Pontiac Motor Division

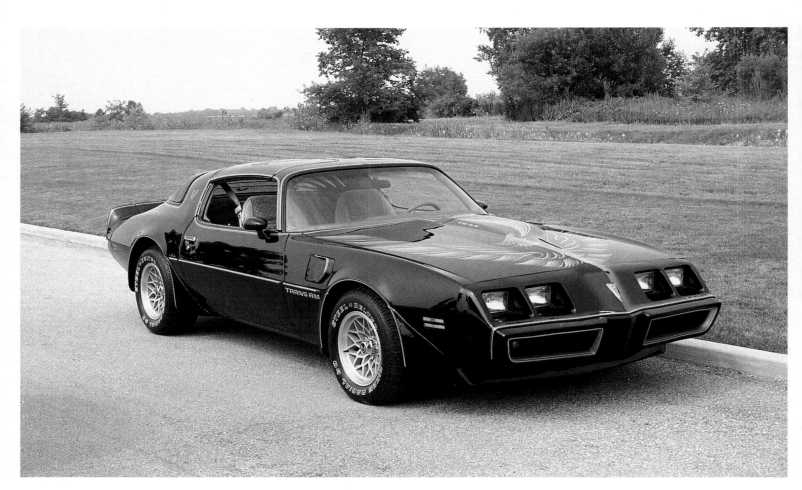

For some reason, the 1980 Trans Am didn't catch on with the buying public, and fewer than half as many as in the previous year were sold. This Special Edition came in black with gold trim, and featured the spoiler, wheel flares, roof edges, and custom striping. A real looker to be sure! The front end of the 1980 Trans Am changed little from the previous year and featured the headlights in four separate boxed and recessed areas. Note that the front wheelwell flares end in a point at the front grille. A number of different powerplants were available, providing the buyer with a choice of economy or performance.

A limited number of the Tenth Anniversary models were garbed as Daytona 500 Pace Cars, making them even more rare and more desirable for collectors.

Moving up to the first year of the new decade, there was devastation in the Firebird production ranks, with sales totals dropping to almost half of those for 1979, with only 107,340 built. The 1980 Trans Am would also be heavily hit, dropping like a rock to only 50,896.

Engine choices were broad for this model year, with five offerings: one V-6; and four V-8s. The base, 3.8 liter (231ci) V-6 (a Buick-built powerplant) kicked out 115nhp and provided peppy performance along with excellent gas mileage.

Alongside the V-6 were three versions of a 4.9 liter (301ci), Pontiac-built powerplant. The Code-W standard version offered 140nhp. Next came the 4.9 E/C Code-Y version that kicked the power up to 155 net ponies, while the top version was the Code-T turbo version, pumping out an impressive 210 horses. The 400 and 403ci engines were gone.

With the turbo-topped version came a stylish special hood featuring an offset "Power Bulge" and "Turbo 4.9" lettering. The engine could be ordered with either the Trans Am or the Formula and was a popular engine choice. Almost half the 23,422 Trans Ams built in 1980 carried the powerplant.

The turbo powerplant used an electric spark control system that helped it run close to peak combustion pressures. The turbo boost was maintained at about 9psi as the turbocharger came up to higher speed, so the turbo didn't over-boost the engine. The compression ratio was listed at 7.5:1 with 345lb-ft of torque.

A number of special option packages was offered in 1980, including a Yellow Bird package for the Esprit which featured a two-toned yellow motif. There was again a Formula appearance package, but most of the special options this year were carried in two special Trans Am editions and an Indy Pace Car replica.

Two versions of the special edition were available, with and without a T-top. The option came in black and gold with all the goodies, such as a spoiler, wheel

There were three versions of the 4.9 liter powerplant available in 1980. The Code W standard version was capable of 140hp. Then came the Code Y version, providing an additional 15hp. But the killer is shown here: the turbo version, kicking out an impressive 210hp. Note the off-center hood blister to clear the new turbo powerplant. This model also carries the classy Turbo cast-aluminum wheels.

This Indy 500 publicity photo shows a sea of Trans Am Pace Cars behind the real pacer and a vintage Indy race car. That's former driver Johnny Parsons, Sr., who drove the pace car that year. The 1980 Trans Am Pace Car remains one of the best-looking pacers in history. Indianapolis Motor Speedway

One of the most popular movies of 1981 was the Burt Reynolds movie, Smokey and the Bandit. This Special Edition model commemorated the movie with gold detailing highlighting the black body. The Special Editions were T-top coupes with the 4.9 turbo powerplant. The front-end treatment for the 1981 Firebirds and Trans Ams was basically unchanged for the model year. The lower, pouty snoot protruded far forward, with the four headlights being individually enclosed. This would be the final year for this front-end design.

flares, roof edges, and custom striping.

But the kingpin for the year was the Indianapolis 500 Pace Car X87 model, a

Previous page
To many, the 1980 Indy Pace Car Trans Am was the best looking pacer ever built. The big bird was sprawled across the hood in gray along with stylish red and charcoal striping. The pacers carried the 4.9 liter turbo powerplant, three-speed automatic transmission, and 3.08:1 rear-axle ratio. Options included a custom interior, large bird hood decal, and the appropriate pace car door lettering.

specially detailed Trans Am with Indy graphics on both sides. There was also a small "Indy 500" logo below the Trans Am lettering on the lower front quarters. The pace cars carried only the 4.9 liter turbo powerplant, and were a popular seller, with approximately 5,600 built.

The pacer was white with a charcoal accent extending from the hood through the upper door, the front of the sail panel, and the forward half of the roof. Red and charcoal striping was the final accent. Silver-tinted, removable hatch-roof panels and aluminum white "air-flow" wheels were standard equipment. Along with the turbo 4.9 liter powerplant, the

pacers toted the three-speed automatic transmission and a 3.08 rear axle.

The big change for 1981 was "Computer Engine Control." The stated goal of the system was to provide better fuel economy and lower exhaust emissions while still maintaining the same level of performance. Not an easy task, but one that Pontiac managed to accomplish.

The new system certainly didn't do anything for showroom activity, as Firebird production declined by more than 36,000 units from 1980's totals.

Powerplant-wise, the model year saw the introduction of a new Code-S 265ci (4.3 liter) V-8. The 305 (5.0 liter) four-

The 1982 Trans Am had a clean, slinky look, with its long, flat rear window, flat rear spoiler, and race-style wheel covers. Pontiac Motor Division

barrel engine was available in several versions, and the 4.9 liter (301ci) would be available in both the E/C and turbo versions, the final year for both.

As in the previous model year, there were the Special Edition coupe and T-top versions (Model X87) of the Trans Am, along with another pace car creation. This time, the 2,000 pacers carried the distinction of pacing the 1981 Daytona 500, the third straight year there had been a Trans Am leading the field. Only the 4.9 Turbo was available with this model.

But the most famous Trans Am mutation of the year, and maybe the most famous of any year, was the so-called "Bandit" model. Inspired by the Burt Reynolds movie *Smokey and the Bandit*, a New Jersey firm modified 200 Trans Ams with a 462ci powerplant, five-speed transmission, and other goodies. These machines carried a monstrous-for-the-time $30,000 price tag.

A New Generation of Birds

You'd have to firmly state that 1982 ushered in a new era in Firebird/Trans Am styling, handling, and performance. The buying public liked what it saw, with a monumental boost in sales to 116,362, of which 52,960 were Trans Ams. The figure was the highest since 1979 and ex-

ceeded figures of all the following years right up to the present.

The body styling was completely new and placed an emphasis on improved aerodynamics. A steeply raked windshield (at a sixty-two degree angle), a low sweeping nose, a squared-off tail (some of which carried an optional rear wing), and a redesigned grille really set this model apart from the crowd. These attractive lines will undoubtedly make these cars desirable for many years.

Handling was also greatly enhanced by the addition of a rear torque arm system, front MacPherson Struts, and a weight shaving down to 2800lb—even lighter than the original Firebird.

Model offerings were down for the 1982 model year to only two Firebirds

87

The 1983 Trans Am was one tough performer, with a 5.0 liter powerplant capable of 190nhp and 240lb-ft of torque. The Trans Am sported few body changes from the previous year. Pontiac Motor Division

(the base model and an S/E Coupe) and the Trans Am coupe. Unlike previous years, there were no special editions. However, some 2,000 Trans Ams were fitted with a special Recaro interior.

A new, four-cylinder, 2.5 liter powerplant with fuel injection provided the base level power. This so-called "Iron Duke" engine is still used with great success in the 1990s in short-track stock car racing. The S/E offered a 2.8 liter V-6, while a well-performing 5.0 liter (which had a dual-throttle-bodied, fresh air version) came with the Trans Am.

Starting with the release of the 1983 models the words "performance" and "power" could again be applied to the Firebird and Trans Am models.

Performance enjoyed a small rebirth in 1983 with four peppy powerplants available. The 2.8 liter V-6 even carried the legendary "HO" designation of earlier days and provided 135nhp.

Four powerplants returned that thrust backward when the pedal was pushed to the floor. The quartet of engines included a 2.5 liter four-banger, a 5.0 liter in both fuel-injected and carbureted versions, and a pair of impressive 2.8 liter V-6s. And surprise! The hotter of the two V-6s carried an almost forgotten "HO" designation. The fun was back—sort of. Production took a dramatic drop to 74,884.

The new HO provided a 135nhp rating, up considerably from 1982's 112nhp version. It came with custom performance pistons, larger valves, and an 8.9:1 compression ratio. Compared to the early days, 135nhp (or approximately 160bhp) sounds pretty feeble, performance-wise. But consider that this V-6 made its power from a mere 173ci compared to the monstrous 400 and 455 powerplants of the past.

The Base Firebird, S/E, and Trans Am models remained basically unchanged, but there was a multitude of Trans Am Special Editions.

Another race commemorative model was the Twenty-Fifth Anniversary Daytona 500 Pace Car. The Recaro Trans Am was a highly detailed, gold and black beauty, requiring an additional $3,610 when the cross-fire-injected, 5.0 liter engine was ordered.

This custom-paint version of the 1984 Trans Am was one classy-looking black and gold beauty. The model featured the gold-colored lower rocker panel extensions with thin, golden, horizontal lines on the bottom of the body. The Trans Am was quickly becoming the showpiece of the pony car world in 1984, and the turbo was still the preferred powerplant.

Even though changes were very minor for the 1984 model year, the production would show a monumental boost to 128,304, thanks in part to the racy looks of the Trans Am of which 55,374 were sold. The same three models—the first level Firebird, S/E, and Trans Am—continued as the tri-model offerings.

Powerplants remained practically unchanged, with the 2.5 liter four-cylinder now carrying a 9:1 compression ratio and a 92nhp rating. A 2.8 liter two-barrel powerplant was also available. The 2.8 liter HO (would you believe?) carried only a two-barrel carb! With an 8.9:1 compression ratio, it was rated at 125nhp. Two 5.0 liter mills were available, although the top L-69 version (150nhp) was available only with the Trans Am.

The base Firebird and the S/E remained fairly sedate, but such was not the case for the Trans Am. Options included the heavy-duty WS6 Suspension Package and five-speed transmission. The Recaro Option was again available, along with the low-mounted, W62 Aero Package and a number of interior luxuries. With the hotter 5.0 liter, this is indeed an excellent find.

Two other special models appeared in 1984. The first was the early model year introduction Fifteenth Anniversary Trans Am, which carried many special appointments including a horizontally-split dark bottom and light top paint scheme, hatch roof, and custom interior. Then there was the Mecham racing creation called the Motor Sports Edition Trans Am. Model identification was carried horizontally across the doors and at the top of the windshield. The numbers of these cars were few, but if one can be found, its a good investment.

Refinements continued for the 1985 Firebird with certain upgrades in perfor-

Nope, you couldn't go out and buy a Trans Am station wagon in 1984. But Pontiac did build this concept car that was displayed around the country.

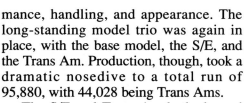

mance, handling, and appearance. The long-standing model trio was again in place, with the base model, the S/E, and the Trans Am. Production, though, took a dramatic nosedive to a total run of 95,880, with 44,028 being Trans Ams.

The S/E and Trans Am both showed considerable facelifting. The S/E had both new front and new rearend treatments, giving the model more of a European look. The S/E identification was carried on the lower doors and the bird decal rode on the sail panels.

But the real changes came to light on the dramatically new Trans Am. This was a "looker" of the first order, with race car inspired styling right off the showroom floor. The W62 ground effects package gave the design a look of speed

The Trans Am design that had been around for several years continued in 1985, but there were some body refinements. Performance improvement was the goal of the 1985 Trans Am, and Pontiac proved it with powerplants with tuned-port injection and Rally tuned suspension. Outside, the aero package had been increased in size, setting off the model in a big way. The "Trans Am" lettering continued to be inscribed on the right side of the rear bumper. Production was down sharply for the year, with only 44,028 being constructed. Pontiac Motor Division

when it was standing still. The aerodynamic add-ons traversed the complete length of the body and greatly reduced the ground clearance. Another nice touch was the swoopy-looking, rear deck-mounted spoiler.

And suddenly, after an absence since 1981, the hood Bird was back, but its dimensions were much smaller than in the past.

As in the previous four years, a Trans Am paced the Daytona 500, but there was no limited edition model to honor the occasion this time.

The punch was back under the hood in 1985, with a 135nhp 2.8 liter V-6 (by Chevy). Three versions of the 5.0 liter engine provided impressive performance. The base version was worth 155bhp, while the HO and TPI (Tuned Port Injec-

tion) versions made 190 and 205 net horses, respectively. Performance was no longer a bad word in 1985.

The following year, the 2.8 liter four-cylinder went away, leaving only the 2.8 liter V-6 (again Chevy-built), which was fuel-injected and again worth 135nhp. The three 5.0 liter engine choices were worth almost the identical power values of the previous year.

The front-end treatment for the 1986 Firebird was slightly changed but retained the same overall look. There was also a new body stripe treatment as well as upgrades to the interior and a new clear coat paint finish.

The first-level Bird got some attention in 1986 with a new rearend treatment and a new body stripe design. For the S/E, there was a number of interior improvements.

The 1986 Trans Am got a new 140mph speedometer and a revised hood bird. The Y99 suspension package incorporated front and rear stabilizer bars along with custom shocks.

Nineteen eighty-seven was the start of a downward production spiral that would continue through the 1991 model year and, in view of the general downturn of the economy in the early 1990s, will probably extend beyond that. But Pontiac was trying hard to please the market and came out with a barrage of new higher horsepower ratings to hopefully generate buyer interest.

The 1987 Trans Am GTA sported big numbers under the hood, with a 350ci (5.7 liter) mill, again built by Chevy, which provided a potent 210 net horses. PMD figured the performance was necessary in order to counter the advertising blitz that Ford had initiated with its 5.0 liter Mustang. Several different axle ratios were available with this powerplant, making it a potent performance offering.

The 1986 Trans Am sported a racy look with the optional Aero package and the rear deck-mounted spoiler. The rear-mounted brakelight was placed on the Firebird and Trans Am for the first time, in 1986. It was a step in the direction of safety for the low-slung models, allowing better visibility from the rear. Two powerplants were available with the model: the 160nhp and 210nhp versions of the 5.0 liter mill. This car belongs to Tom Johnson of Dayton, Ohio.

Next page
The GTA was Pontiac's bird of prey for the Firebird line in 1987. Besides the great looks there was also a 350ci powerplant under the hood, kicking out 210nhp. The Pontiac buyer could also acquire a number of different rearend ratios to suit his or her driving style. The battle was on with the 5.0 liter Mustangs.

PMD liked to call the 1988 model year "a refinement of the strengths established in 1987." The "Fun to Drive" Firebird, "Street Performer" Formula (which returned after a several-years' absence), "Muscular, High-Styled" Trans Am, and "The Ultimate" GTA were the factory descriptions of these models.

The 5.7 liter Tuned Port Injection mill was still at the top of the heap, putting out 235 net horses with 3.27 and 2.77 rearend gearing available. The powerplant was available with two Trans Am models and with the Formula. Three 5.0 liter engines demonstrated impressive power options at 170; 190 with an automatic and 215 horses with the five-speed manual transmission.

The GTA was a model in itself for 1988, elevating it above the rest of the Firebird line. It featured a multitude of options, including the WS6 Performance Suspension Package: tuned springs, gas shocks and struts, and four-wheel disc brakes. Just over 20,000 were built.

The Formula nametag showcased a number of optional features including all

The Formula was one tough-looking model for 1988, as demonstrated by this Fire Engine Red machine. Top powerplant for the Formula was the 5.7 liter Tuned Port Injection engine putting out a potent 235nhp and *your choice of a pair of rearend ratios. There was no doubting its identification with the large "Formula" lettering on the lower portion of the doors. The Formula got the WS6 package in addition to the 16in alu-* *minum deep-dish wheels. On the dash, there was also a 120mph speedometer and a tachometer.*

The power was back with the top gun power-plant for 1988 being this 5.7 liter, 350ci powerplant providing 210nhp. Numerous powertrain options were available to tailor the power to the buyer's driving style. The engine didn't have the horsepower figures of the good old days, but with modern technology, the performance was almost equal to the earlier models'.

The GTA was the magnificent top-of-the-line Firebird in 1988. It was loaded to the hilt with gas shocks and struts, four wheel disc brakes, a high-performance suspension package, and tuned springs. Suddenly, performance was back in vogue.

The top sporty model of the 1988 Formula was the Formula 350, with lettering that stretched across both doors. The model was also accentuated by the complete lower body blackout treatment. With the aluminum wheels and rear spoiler, this was one of the best-looking sports cars ever built.

of the big horsepower options. It got the WS6 package along with the new, 16in, cast aluminum, deep dish wheels. Inside, there was a 120mph speedometer with tach and gauges. The standard Firebird model carried the 2.8 liter, fuel-injected V-6 as standard equipment.

Does the Trans Am ever get tired of being a pace car? Apparently not! It happened again in 1989 with the Indy honor again being bestowed upon it. But, national exposure or not, Trans Am production fell to approximately one-fourth its 1988 level, with only 5,727 sold.

If there was a Bandit I, there would most likely be a Bandit II, and it arrived on the scene in 1988. The specially detailed machine carried the Bandit II identification.

The 1989 model year was the twentieth (didn't seem that long!) anniversary of the Trans Am. And, of course, there had to be a 20th Anniversary model to celebrate. Would you believe that even the spark plug wires got the appropriate lettering!

Normally it's necessary to build a special engine to power the actual Indy pace car. That was not the case here, as the 3.8 liter Buick V-6 was more than up to the job. The turbocharged powerplant was capable of an impressive 245hp. These cars will continue to be interesting collectibles in the years to come.

The horsepower numbers of the Pace Car Edition (there were only 1,555 built) were up to an impressive 245nhp from a 3.8 liter, turbo charged, Buick V-6 engine hooked to a four-speed automatic.

Previous page
The 1989 Indy 500 was the second Indy Pace Car honor for the Trans Am line. Just 1,500 pace car replicas were built with a 245hp version of the 3.8 liter Buick V-6 under the hood. Performance was advertised as 13.8sec in the quarter-mile and a 150mph top speed. The pace car edition was one beautiful machine, and there were hardly any differences between it and the real pacer. The model carried a number of options including these gold-tinted, honeycomb wheels.

The pace car replicas were almost identical to the real pacers externally, missing only the strobe lights of the race day models. Performance-wise, the model didn't have to be modified for its pacing duties. The performance was advertised as a 13.8sec quarter capability and a 150mph top speed.

There wasn't much happening change-wise in the Firebird lineup for 1990, just minor deviations from the pre-

The Firebird and Trans Am continue to be popular vehicles for officers of the law. This Ohio county sheriff's machine is a 1989 Firebird and is out there supporting the laws of the land. That light bar really sets off the car in a different way.

vious year. The Trans Am really looked as if it had fallen from favor with the buying public, with only 2,507 (including GTAs) built. It was the lowest TA total since 1971. Total Firebird production was also a minimal 20,453.

But things turned around for 1991 when Trans Am production more than doubled and Firebird production was up to 43,306. The increase, though, was not difficult to understand, considering all the upgrades that were introduced. The exterior was completely changed with restyled fascias front and rear. The low profile Endura front end included an integral air dam. The GTA, Trans Am, and Formula all received a redesigned rear deck spoiler.

The 1991 Firebird and Trans Am were largely revamped and the completely new design set the industry on its ear. The buying public liked what it saw, and production showed a considerable increase for the model year. The grille design was probably the area of greatest change for the 1991 model. The design looked back to the first- and second-generation Firebird styling with the twin grille openings, but obviously the old style never looked like this. The teardrop openings were truly unique, to say the least.

Again, the GTA was the big gun in the Firebird line for 1990. The longstanding body design had been modernized just about as far as it could go; it would be the final year of the body design that had held forth during the 1980s. The model, though, seemed to have fallen out of favor, as only 2,496 were built for 1990. These cars are difficult to locate even though it's only been a few years since they were produced. Larry Warf

106

The 1991 model year marked the return of a convertible to the Firebird line-up for the first time since 1969. The top was manually operated and offered for both the Firebird and the Trans Am; base prices were $19,159 and $22,980 respectively. Pontiac Motor Division

The hot new items included convertible options on all models; a new sport appearance package; a new spoiler for

the Formula, Trans Am, and GTA; and a new aero treatment for the Trans Am and GTA. New, 16in, charcoal Diamond Spoke wheels also came with the Trans Am.

As in the past, the GTA was again the top of the line and featured the 240hp, 5.7 liter, TPI powerplant and four-speed transmission. Two TPI 5.0 liter motors were also available, producing 230 and 205nhp, respectively. The same engines were available with the Trans Am.

A 170nhp version of the 5.0 liter came standard with the Formula, with the aforementioned mills also offered as options for the standard Firebird. The standard powerplant was the 3.1 liter, multiport, fuel-injected V-6.

The distinctive GTA aero package carried fog lights and brake cooling ducts in the front fascia, distinctive new side aero treatment, functional hood louvers; and air extractors. The WS6 sport suspension and limited slip rearend were

The interior of the 1991 model was upgraded right along with the exterior. The design combined both luxury and dashing sportscar looks to produce this great cockpit.

Power was back in fashion with the new body design for the 1991 model. The 5.7 liter TPI turbo powerplant was worth 240nhp that could really put this machine down the road.

In mid-model-year 1991, Pontiac unveiled a new convertible model Firebird and Trans Am. It was the first Firebird convertible since the 1969 first-generation cars passed into history. And the styling of the convertible was glorious either in top-up or top-down position. There were minimal changes incorporated in the 1992 convertible. But after what had been done in the previous year, that was understandable. Two engine choices provided 160nhp from a 3.4 liter V-6 and 280nhp from a 5.7 liter V-8.

During the late 1980s and early 1990s, Pontiac built a number of prototype cars and engine systems. This interesting engine was a 350ci TPI engine with prototype twin turbos. Its performance might have been outstanding but we'll probably never know.

Pontiac Motor Division

It was called a street car, but the 1992 special edition Formula Firehawk was a race car in every sense of the word. It didn't look that much different from the normal version, but that certainly wasn't the case inside. The Firehawk Competition Option was created by SLP, which stood for Street Legal Performance Engineering, and was denoted as RPO B4U. It included a Corvette ZF six-speed gearbox, stainless-steel exhaust header and dual catalytic converter setup, a tuned-length downdraft inlet manifold system, four-piston Brembo cross-drilled F-40 front disc brakes, Recaro driver's seat, Simpson five-point harness, rear seat delete, and a lightweight hood. With its specially prepared powertrain and suspension, the Firehawk was a solid 13sec quarter-mile performer.

standard. The Formula was quite recognizable by its distinctive power bulge hood and specific Formula graphics.

And finally, in 1992, it was time to celebrate the twenty-fifth anniversary of the Firebird. The model had stood the test of time and continued to stand tall, though it must be admitted that it was probably still in the shadow of the Camaro.

Things remained basically unchanged under the hood for the 1992 model year with the 5.7 liter TPI V-8 again providing its substantial 240 net horses. There was also a pair of 230 and 205 nhp 5.0 liter engines and a 140hp V-6.

The biggest improvements for the 1992 model, though, were invisible to the eye. Stronger welds and an increase in noise-softening adhesives were now in place internally. Externally, there were new body graphics.

Undoubtedly 1992's most exciting model option was the Firehawk. These

This was the 1992 Formula Firehawk's specially prepared 5.7 liter mill that was capable of pumping out 350hp. It carried a four-bolt main, forged crankshaft, aluminum pistons, and stainless-steel headers. It was better than the good old days with its 350hp and 390lb-ft of pavement-scorching torque. The 0-60mph performance was an awe-inspiring 4.9sec. The Firehawk was a performer of the top order, one of the hottest machines ever produced in this country.

highly modified Firebirds possessed performance approaching that of a ZR1 Corvette. Firehawks were priced at $40,000, but could cost as much as $50,000 if the competition version was desired. The additional $10,000 purchased such items as huge front brakes, Recaro seats, Simpson five-point harness, deleted rear seat, and an aluminum hood.

A complete redesign highlighted the 1993 Firebird, with over ninety percent of the parts and pieces being new for the

Next page
The styling of the new 1993 Firebird line was kept under cover until the official debut. This was the prototype model, shown going through the paces on a General Motors test track. Pontiac Motor Division

The Firebird Coupe was all new for the 1993 model year. Pontiac targeted the model at the college-educated singles set. The model was designed to compete against the Ford Mustang LX, Dodge Daytona ES, and Dodge Stealth. Pontiac Motor Division

model year. Ride and handling were improved by a new Short and Long Arm (SLA) type front suspension which incorporated high-pressure monotube shocks.

Yearly Sales History	
1967	74,205
1968	96510
1969	69,136
1970	57,230
1971	49,078
1972	31,204
1973	43,669
1974	75,565
1975	98,405
1977	133,154
* 1978	175,607
1979	170,497
1980	95,449
1981	61,460
1982	83,810
1984	105,628
1985	101,797
1986	96,208
1987	77,635
1988	58,459
1989	60,267
1990	39,781
1991	24,601
** 1993	9,679

*All time sales record.
**Through May.

Next page
The Formula was the next step up for the Firebird line in 1993. The model featured body-colored sport mirrors, smoothly contoured taillights, and 16in silver sport aluminum wheels. Pontiac Motor Division

The awesome 275nhp 5.7 liter powerplant was standard on the 1993 Trans Am. The powerplant was mated to a six-speed manual transmission, although a four-speed automatic was also available. The performance package included performance-calibrated suspension with a 3.23:1 axle ratio, engine oil cooler, and a 155mph speedometer. Pontiac Motor Division

Two engines were available: the 160hp, 3.4 liter V-6 (L32); and a 280hp, 5.7 liter V-8 (LT-1). Added safety equipment included standard Delco Moraine ABS VI brakes and standard driver and front passenger airbags. The Pass-Key II Theft Deterrent system is also standard on all models.

Firehawks were again offered in 1993. The fuel injected 350 LT-1 engine produced 300nhp and was mated with either a six-speed manual or a four-speed automatic transmission. Performance was awesome, with quarter-mile times of 13.5sec at 103.5mph.

For 1994, two Firebirds (base and Formula versions) and two Trans Ams are offered. The highly advertised Formula comes equipped with the 275bhp LT1 350ci engine mated to a six-speed manual transmission. The base model carries the 160nhp L32 207ci V-6.

Handling duties on the Formula are accomplished by the same FE2 suspension system fitted to the Trans Am. This includes gas shocks and front and rear stabilizer bars. Stopping is performed by four-wheel disc brakes.

Pontiac designed the Formula to do sales battle with such sportsters as the

Firebird Production Figures		
Model Year	Trans Am Production	Production Total
1967	0	82,560
1968	0	107,112
1969	697	87,011
1970	3,196	48,739
1971	2,116	53,124
1972	1,286	29,951
1973	4,802	46,313
1974	10,166	73,729
1975	27,274	84,063
1976	46,701	110,775
1977	68,745	155,736
1978	93,341	187,284
1979	117,108	211,454
1980	50,896	107,340
1981	33,492	70,899
1982	52,890	116,362
1983	31,930	74,884
1984	55,374	128,304
1985	44,028	95,880
1986	48,870	110,463
1987	32,890	88,612
1988	20,007	62,455
1989	5,358	64,404
1990	2,496	20,453
1991	5,353	33,832
1992	2,151	27,567

The large analog gauges were easy to read. The tachometer and trip odometer were clustered in the driver's direct line of sight, while the sound system and HVAC controls could be reached without the driver ever having to lift his or her back from the seat. Pontiac Motor Division

Ford Probe GT, Mustang LX, Dodge Daytona IROC RT, and the Dodge Stealth ES.

Carrying a lower sticker price than in 1993, the 1994 Trans Am's two versions (standard and GT) differ largely in what options are fitted as standard. Both versions have the 275nhp 350ci LT1 engine and six-speed manual transmission. The Trans Am's new look includes integral fog lamps and rocker panel extensions. At the rear is a sleek, integrated rear deck lid spoiler.

Through nearly three decades, the Firebird has had an interesting history, rolling with the motoring trends, but occasionally setting its own trends.

The Firebird, along with its upscale derivatives the Formula and Trans Am, experienced the muscle era of the late 1960s and early 1970s in all its glory. And with the fabulous Super Duty 455, it actually outlived most of its muscle car peers.

In the 1980s, the Firebird was a leader in the performance resurgence. This role continues into the 1990s with such awesome models as the Firehawk.

Besides performance, the Firebird has always had good looks in its favor. Over the years, there have been a number of special editions, anniversary models, and pace cars; all of these continue to attract collectors in the 1990s.

The Firebird will continue to be revered for its performance, design, and history . . . and don't ever confuse it with a Camaro!

Racing

Although it's not well documented, the Firebird and Trans Am have each been involved in both Trans-Am road racing and NHRA/IHRA drag racing.

The models have never been a part of the NASCAR Winston Cup scene, as that circuit uses only full-bodied sedans, but the models have been heavy players in the national short track circuits, including ASA, All-Pro, ARTGO, and ACT.

Trans-Am Road Racing

Unlike Chevy (with the Z28 Camaro model), Ford (with the Boss 302), and Chrysler (with the T/A Challenger), Pontiac did not introduce a special model to publicize its Trans Am racing participation.

Pontiac initiated a low-level Trans-Am effort in 1967 with a Firebird racer powered by a 230ci V-6 topped by three single-barrel Weber carbs. It was a curious choice of powerplants in a land of powerful V-8s, and it achieved little success. The model was never produced in a street version.

Pro Stock driver Jerry Eckman has long been known for his association with Pennzoil-sponsored Trans Am drag machines. The machines have been competitive in National Hot Rod Association (NHRA) competition.

Pontiac has run the SCCA Trans Am series for many years. But in 1991, the Firebird reappeared after an absence since 1984 with *this EDS-sponsored effort. This promotional release shows the paint scheme of the car.* Pontiac Motor Division

However, the powerplant for one 1968 Trans-Am Firebird used a Chevy Z28 engine which had been built in Canada. It was a perfectly legal situation for Trans-Am qualification, since the engine was manufactured out of the country. The use of this engine was required at the time because Pontiac had no small block V-8 of their own.

There was also a short-term experiment in the late 1960s when Pontiac decided to do a significant destroke of its 400ci powerplant to reduce it to 304ci. This powerplant was fitted with high-compression Ram Air heads and two four-barrel carbs. The mill reportedly made over 475hp, well in excess of the Camaro competition. Unfortunately, this awesome powerplant never saw production for street versions, but it certainly proved to be a potent performer on the track.

Pontiac's early Firebird Trans-Am racer weighed about 3350lb. Suspension duties were handled by independent upper and lower control arms and coil springs at the front, and a live rear axle and leaf springs at the back.

The Firebird team played a role in the points finishes of two drivers in the 1968 season. The first was Jerry Titus, who used both a Mustang and a Firebird (which he only drove once) to finish third in the Trans-Am series.

The second driver was Craig Fisher who did a better job of showing the Firebird's capabilities. He finished fourth in the series, garnering most of his points in

Firebirds and Trans Ams have long been stalwarts in short-track racing throughout the United States. Here, Mike Eddy—long a superstar with the American Speed Association (ASA)—runs a Firebird during the 1985 *season. The organization required that the builders conform closely to Pontiac body lines, and Eddy's racer actually checked with factory templates.*

Previous page

Veteran driver Joe Shear has been a successful user of Firebird-based, short-track stock cars. He is shown in action with the American Speed Association in 1985. The aerodynamics of the Firebird made it an effective performer on the nation's short tracks.

a Firebird, though he did drive a few races in a Camaro. His Firebird finishes included two seconds, one third, and two fourths. It was a great season for Fisher, no doubt, but series winner Mark Donohue had won ten races to take the title, a

performance that was pretty tough to top.

Titus ran the whole 1969 season in a Firebird, finishing third in the points with one win, three thirds, two fourths, and a fifth. Milt Minter brought home a tenth place points finish with two fifths and a pair of tenths. In 1972, Minter did himself and Pontiac proud with a runner-up finish in the points.

The Firebirds running during the 1970 to 1972 period used a 305ci engine with a single four-barrel and 450-plus horsepower. The cars carried four-speed transmissions, forged steel spindles, and four-wheel disc brakes.

There was a two decade gap before a

PMD logo car would show up again in the points race. Darin Brassfield finished fifth in the points in 1983 with two fourths, one fifth, and a tenth.

Nineteen eighty-three was a great year for Pontiac, with Elliott Forbes-Robinson ending up fourth in the points, followed by Frank Leary in fifth, both driving Trans Ams. Pontiac's best finish recently in the Trans-Am series was in 1985: Bob Lobenberg ended up fifth in the points, Jim Miller seventh, and Jim Derhaag tenth.

But probably more important than the victories themselves to the companies in the Trans-Am series were the placings in

The Firebird and Trans Am have played heavily in short-track racing through the *years. Here, long-time Pontiac runner Dick Trickle carried one of the many checkered* *flags he won in ARTGO short-track competition in the late 1970s.* **ARTCO** *photo*

the manufacturers' points. For the Firebird and Trans Am (both of which counted for Pontiac points), there were some significant company accomplishments.

In 1982, PMD actually won the Manufacturers Championship and was second a year earlier, with third place finishes accomplished in 1969, 1972, 1978, 1984, and 1985.

Short Track Racing

For many years, the Firebird and Trans Am have been vigorous performers in short track racing in both the United States and Canada. In fact, during the 1970s and into the 1980s, many circuits were almost completely Camaro- and Firebird-dominated. When a number of competing body styles, such as Luminas and Thunderbirds, were introduced in the mid-1980s the Pontiac models still remained solid contenders.

Most of these short track machines wore fiberglass replicas of the Firebird/ Trans Am body shapes, and with the exception of fender flares and wider wheel cutouts for the wide racing tires, the racers were very close to stock contours. Either a 350ci V-8 or a smaller V-6 provided the power for these machines. The Firebird and Trans Am models have been competed in such national series as ARTGO, ASA, ACT, All-Pro, and others.

The big name with Pontiac during the 1970s was current NASCAR driver Dick Trickle. Racing on the ARTGO circuit, Trickle turned to Firebirds in 1977 and ran them until 1980, during which time he brought the PMD logo to the checkered flag twenty-one times. There was also a string of seven consecutive wins in 1979, with Trickle taking the ARTGO championship in both 1977 and 1979. During the 1980s, Trickle raced a mutation combining both Firebird and Camaro elements, namely a Camaro nose section and a Firebird tail.

Drag Racing

The Firebird has also played in the national drag racing scene, with participation in both the NHRA and IHRA circuits.

NHRA participation in the early 1990s has included racers campaigned by Rickie Smith, Jerry Eckman, and Jim Yates in Pro Stock competition, and Mike Ferderer in Super Comp. During IHRA Pro Stock competition, Harold Denton turned a blazing 6.989sec run at 198.18mph at the 1991 US Open Nationals in a 1991 Trans Am.

In the bracket classes at hundreds of local drag strips across the country, you will see new and old Firebirds and Trans Ams blasting down the strip. They are two of the most popular contenders at this level of Saturday night hometown racing.

Speed Records

In an amazing 1991 speed effort, Dave Macdonald pushed his 1991 Trans Am to 268.799mph to set the A-Gas Coupe speed mark at the Bonneville Salt Flats. If mechanical problems had not been encountered, there could well have been a 300mph run. Macdonald's Trans Am also shattered the nitrous category record with a 272.203mph clocking. Macdonald indicated that the Trans Am body style was aerodynamically superior to any other available. "Since I wasn't allowed to change the body lines any, the car ran just the way it came from the showroom and it performed extremely well," Macdonald revealed.

Above and right
Drag racing abounds with Trans Am-style drag machines. Shown here are the Top Alcohol Funny Car of Randy Anderson and the Pro Stock racer of Rickie Smith. Pontiac Motor Division

Index